AF686657

AGRICULTURAL SUSTAINABILITY

AGRICULTURAL SUSTAINABILITY

Elisa Gomez Gonzalez

Delve
PUBLISHING

www.delvepublishing.com

Agricultural Sustainability

Elisa Gomez Gonzalez

Delve Publishing
2010 Winston Park Drive,
2nd Floor
Oakville, ON L6H 5R7
Canada
www.delvepublishing.com
Tel: 001-289-291-7705
001-905-616-2116
Fax: 001-289-291-7601
Email: orders@arclereducation.com

© 2019 Delve Publishing

ISBN: 978-1-77361-440-3 (Hardcover)

This book contains information obtained from highly regarded resources. Reprinted material sources are indicated and copyright remains with the original owners. Copyright for images and other graphics remains with the original owners as indicated. A Wide variety of references are listed. Reasonable efforts have been made to publish reliable data. Authors or Editors or Publishers are not responsible for the accuracy of the information in the published chapters or consequences of their use. The publisher assumes no responsibility for any damage or grievance to the persons or property arising out of the use of any materials, instructions, methods or thoughts in the book. The authors or editors and the publisher have attempted to trace the copyright holders of all material reproduced in this publication and apologize to copyright holders if permission has not been obtained. If any copyright holder has not been acknowledged, please write to us so we may rectify.

Notice: Registered trademark of products or corporate names are used only for explanation and identification without intent of infringement.

Delve Publishing publishes wide variety of books and eBooks. For more information about Delve Publishing and its products, visit our website at www.delvepublishing.com.

ABOUT THE AUTHOR

Elisa Gomez was born and raised in Canary Islands; her personal mission is to incorporate Corporate Social Responsibility and Sustainability in all businesses. She strongly believes in the positive impact and added value it brings to the business itself and to all.

After over 6 years in the corporate world, she founded better together, a sustainability consulting and advisory company that enables all types of business access to sustainability management. Their aim is to provide businesses with the necessary tools to identify calculate and maximize the positive impact they have upon their sphere of influence. Their mission is to spread sustainability, raise awareness and collaborate with change makers into a more just, understanding, respectful and inclusive world.

With a Master's degree in International Corporate Social Responsibility and with more than 7 years of experience as Sustainability expert in International and Startups companies from a variety of sectors, Elisa is an expert in efficient and impactful Sustainability: Strategy definition, Implementation, Analysis, Training and/or Reporting.

Sustainability management can be simple, effective, understandable and affordable! We all have to play our part! Sustainability made simple for students, professors and business, aims to provide to all a comprehensive approach to sustainability implementation, focus on inclusion and raise awareness.

TABLE OF CONTENTS

List of Figures .... *xi*

List of Tables .... *xv*

Preface .... *xvii*

Chapter 1 An Overview on Agricultural Sustainability 1

1.1. Introduction 2

1.2. Definitions of Sustainability 2

1.3. Agricultural Sustainability Strategy 3

1.4. Past Studies on Agricultural Sustainability 4

1.5. Social Dimension Issues on Agricultural Sustainability 6

1.6. Measurement of Agricultural Sustainability 9

1.7. Strategic Interventions Towards Agricultural Sustainability 13

1.8. Impacts of Biotechnology on Agricultural Sustainability 17

Chapter 2 Factor Influencing Agricultural Sustainability 25

2.1. Introduction 26

2.2. Resources 27

2.3. Climatic Factors 27

2.4. Financial Factors 28

2.5. Land Fragmentation 30

2.6. Technological Factors 33

2.7. Information Access 36

2.8. Social-Cultural Factors 37

2.9. Region Factors 38

2.10. Cropping Pattern 39

2.11. Storage Facilities 40

2.12. Agricultural Cooperatives 41

2.13. Policy Formulation 42

2.14. Marketing And Pricing ... 42
2.15. Agricultural Extension Services ... 43

Chapter 3 Irrigation Practices and Agricultural Sustainability ... 47
3.1. Introduction ... 48
3.2. Place of Irrigation in Agricultural Sustainability ... 48
3.3. History of Irrigation ... 52
3.4. Methods of Irrigation ... 54
3.5. Some Factors that Affect Crops Response to Irrigation ... 58
3.6. Problems Facing Irrigation Schemes In Agricultural Sustainability ... 59
3.7. Key Concepts of Irrigation for Agricultural Sustainability ... 61
3.8. Socio-Economic Aspects of Irrigation ... 75

Chapter 4 Cultural Farm Practices That Contribute Toward Agricultural Sustainability ... 79
4.1. Introduction ... 80
4.2. Crops Support ... 80
4.3. Crops Spacing ... 83
4.4. Crops Pruning ... 84
4.5. Mixed Cropping ... 86
4.6. Mulching ... 90
4.7. Crops Fertilization ... 102

Chapter 5 Sustainability of Urban Agriculture ... 107
5.1. Introduction ... 108
5.2. A Brief History of Urban Agriculture ... 108
5.3. Urban Farming for Sustainable Cities ... 109
5.4. The Roles of Urban Agriculture ... 112
5.5. Key Resources for Urban Agriculture ... 120
5.6. Challenges in Sustainable Urban Agriculture ... 129
5.7. Irrigation in Urban Agriculture ... 134
5.8. Wastewater Reuse in Urban Agriculture ... 135
5.9. Solid Waste Management and Urban Agriculture ... 138
5.10. Factors That Determine Land Use For Urban Agriculture ... 140

Chapter 6 **Dairy Cattle Keeping and Agricultural Sustainability 149**
6.1. Introduction 150
6.2. Major Dairy Cattle Breeds 150
6.3. Sustainable Dairy Farming Practices 155
6.4. The Importance of the Dairy Industry 157
6.5. Milk Marketing 158
6.6. Factors Influencing Milk Production in Dairy Cattle 160
6.7. Incentives for Sustainable Dairy Farming 164

Chapter 7 **Beef Cattle Farming and Agricultural Sustainability 167**
7.1. Introduction 168
7.2. Beef Meat Demand and Supply 168
7.3. Challenges in Beef Cattle Production 170
7.4. Markets for Beef Cattle 171
7.5. Factors Affecting Beef Farming 175

Chapter 8 **Sustainability of Poultry, Rabbit, and Fish Farming 187**
8.1. Poultry Farming 188
8.2. Rabbit Farming 194
8.3. Fish Farming 203

Chapter 9 **Arid and Semi-Arid Areas Potentials in Agricultural Sustainability 213**
9.1. Introduction 214
9.2. Major Agricultural Potential In Arid And Semi-Arid Areas 215
9.3. Key Constraints to Agricultural Production in Arid and Semi-Arid Areas 217
9.4. Key Strategies in Tapping Arid and Semi-Arid Agricultural Potentials 219
9.5. Food Security in Arid and Semi-Arid Areas 225
9.6. Solutions Towards Agricultural Food Sustainability in Arid and Semi-Arid Areas 225
9.7. The Factors Influencing Success and Failure in Agriculture 227

Chapter 10 **Organic Farming and Agricultural Sustainability 231**
10.1. Introduction 232
10.2. Sustainability of Organic Farming 232
10.3. The Start of Organic Farming 233

10.4. Elements of Organic Farming 235
10.5. Importance of Vertical Integration In Sustainable Organic Farming 236
10.6. Advantages of Organic Farming 237
10.7. Challenges In Organic Farming 238

References 241

Index 251

LIST OF FIGURES

Figure 1.1. Illustrating key elements of agricultural sustainability.

Figure 1.2. Source of agricultural information.

Figure 1.3. Illustrating the interaction between biotechnology innovation, prevailing conditions and effects on agriculture practices.

Figure 2.1. Illustrating impacts of climate change on agricultural sustainability (it is estimated that 10% of major crops will experience a stagnant or reduced growth rate by 2030 due to climate change).

Figure 2.2. Sustainable agriculture is a key for economic development; the sector requires more financial assistance (capital and capital investment) in order to meet all requirements of population with higher purchasing power.

Figure 2.3. Land fragmentation is a key factor influencing agricultural sustainability.

Figure 2.4. Illustrating an example of technology application in agricultural sector "(A) Increase the productivity using nanopesticides and nanofertilizers; (B) Improve the quality of the soil using nanozeolites and hydrogels; (C) Stimulate plant growth using nanomaterials (SiO_2, TiO_2, and carbon nanotubes); (D) Provide smart monitoring using nanosensors by wireless communication devices."

Figure 3.1. Illustrating irrigation canal systems along the Nile River in Egypt.

Figure 3.2. Illustration exploration of groundwater and its utilization in irrigation schemes for sustainable agricultural practices.

Figure 3.3. Diversion Channel from mountain streams to shallow small channels for irrigation purposes.

Figure 3.4. An example of flood irrigation scheme.

Figure 3.5. An example of row or furrow irrigation.

Figure 3.6. An example of basin flooding irrigation scheme.

Figure 3.7. Soil, water, and salt distribution in soil with a drip irrigation system.

Figure 3.8. Illustrating the effect of evapotranspiration on irrigation practices.

Figure 3.9. The concept of water availability for field crops.

Figure 3.10. Mechanism of plant water absorption.

Figure 3.11. Steps used in analyzing moisture content using gravimetric method.

Figure 3.12. Practical application of electrical resistance blocks in soil moisture content determination.

Figure 3.13. Illustrating the design and application of tensiometer in the determination of soil moisture content.

Figure 3.14. Depicting the crop stages of tomato for water management.

Figure 3.15. The difference in water content between field capacity and permanent wilting point.

Figure 3.16. Illustrating the effects of salt on plants' growth.

Figure 3.17. Illustrating a well-managed irrigation scheme.

Figure 4.1. Illustrating vertical trellis in tomatoes production.

Figure 4.2. Horizontal trellis in tomatoes production.

Figure 4.3. Demonstrating the effect of spacing in maize crop production.

Figure 4.4. Pruning exercise in crop production.

Figure 4.5. Mixed farming practices.

Figure 4.6. Organic mulching.

Figure 4.7. Plastic mulching.

Figure 4.8. Illustrating the effect of mulching on water retention in the soil.

Figure 4.9. Both figure (a) and (b) illustrate delayed weed emergence in a mulched piece of land. The mulches provide favorable conditions for crop growth.

Figure 4.10. Effect of mulching on soil fertility.

Figure 4.11. High humus content enhances the population of microorganisms in the soil.

Figure 4.12. Application of fertilizer on field crops.

Figure 5.1. A practical example of livestock keeping and crop growing in a typical urban area.

Figure 5.2. Illustrating three key pillars of food security as of WHO.

Figure 5.3. Illustrating food waste composting and vertical garden in the city.

Figure 5.4. Illustrating characteristics of green and brown environmental.

Figure 5.5. Illustrating a vacant space under urban farming.

Figure 5.6. A group of youth who have converted a dumpsite into a vegetable farm near Nairobi city in Kenya.

Figure 5.7. Zoning of urban spaces for livestock keeping.

Figure 5.8. Illustrating young farmer growing arrowroot near a polluted river (although the crops benefit from nutrients in the water, this is likely to cause some health risks to the consumers).

Figure 5.9. Showing major sources for urban agriculture finances.

Figure 5.10. Illustrating a group of men with their goat along Nairobi urban areas.

Figure 5.11. Illustrating a group of women engaged in urban agriculture.

Figure 5.12. Illustrating where on urban farm irrigation is greatest by country.

Figure 5.13. Illustrating a farmer in Mezquital Valley, Mexico using treated wastewater in his maize farm.

Figure 5.14. An urban farming system using solid waste from the nearby households.

Figure 5.15. Famers in India urban area taking advantage of vermicompost in their farming.

Figure 6.1. Holstein breed.

Figure 6.2. Jersey dairy cow.

Figure 6.3. Brown Swiss dairy cow.

Figure 6.4. Guernsey dairy cow.

Figure 6.5. Ayrshire dairy breed.

Figure 6.6. Milking Shorthorn dairy breed.

Figure 6.7. Illustrating dairy cows been reared under free stall system.

Figure 6.8. Farmers grinding maize stalks for animals.

Figure 7.1. A typical rural butcher.

Figure 7.2. An example of local government market for beef meat.

Figure 7.3. An example of ranch market.

Figure 7.4. The demand for meat in the world.

Figure 7.5. A cow showing signs of pleuro-pneumonia.

Figure 7.6. Anemic and weakness are some of the common signs of trypanosomiasis in beef cattle.

Figure 7.7. A beef cattle suffering from East Coast Fever.

Figure 7.8. Foot and mouth disease.

Figure 7.9. Cattle trekking for long distance in search of water.

Figure 8.1. An extensive rabbit rearing system.

Figure 8.2. Illustrating a semi-intensive rabbit rearing system.

Figure 8.3. An intensive rabbit rearing system.

Figure 8.4. An intensive fish farming system.

Figure 8.5. Intensive fish farming system.

Figure 8.6. Integrated fish farming system.

Figure 9.1. Livestock keeping in arid and semi-arid areas.

Figure 9.2. Vegetable production in arid and semi-arid areas.

Figure 9.3. Two key water conservation technologies used in arid and semi-arid areas.

Figure 9.4. Beekeeping in arid and semi-arid areas.

Figure 9.5. Irrigation schemes in arid and semi-arid areas.

Figure 9.6. Illustrating some of the achievement Israel has achieved in terms of revolutionizing desert land.

Figure 9.7. A farmer using an oxen plow system to prepare his land for crops production.

Figure 10.1. World scenario of organic farming.

Figure 10.2. Key components of organic farming.

LIST OF TABLES

Table 3.1. Leading countries in irrigated land.

Table 3.2. Outlining the advantages and disadvantages of groundwater in irrigation schemes.

Table 5.1. The number of households engaged in urban agriculture in some of the African cities.

Table 5.2: Access to urban and rural land for crop production for selected towns (as % of total households).

Table 9.1. Agricultural potential concept in arid and semi-arid areas.

PREFACE

The population of the world by the year 2050 is projected to be 9.2 billion. The agriculture sector and most specifically food production are generating and contributing to alarming negative impacts on a social, environmental, and economic level. This calls for an increase in global sustainable agricultural food production system in order to feed the rapidly increasing human population. Food production globally is affected by a series of challenges like salinity, drought, excess heat, floods, nutrient loss, radiation, and anaerobic conditions among others. These contribute to about 30–60% yield decline every year representing some of the major environmental issues affecting agricultural productivity and sustainability.

Agricultural sustainability is a concept that is often interpreted "as the ability of farmland to produce food and other agricultural products to satisfy human needs indefinitely as well as having sustainable impacts on the broader environment." It focuses on agricultural efficiency, environmental stewardship, farm profitability and societal welfare objectives as well as consumer health.

More attention needs to be given to defining key indicators of agricultural sustainability, relating and analyzing these measures to establish trends in agricultural productivity. This analysis must be sufficiently disaggregated and for a long enough time period to allow for spatial and temporal variability inherent in agricultural production.

In the last few decades, the agriculture sector has experienced improvements and advances in numerous areas helping livestock management and crops production. The main advances are: improved fertilizer application, modern irrigation schemes, the use of agricultural machineries, pesticides application, and proper land management. Even though these specific changes have improved significantly the industry their effectiveness for the future is uncertain.

Researchers have suggested the need for new innovative approaches that will help towards the integration of ecological and biological processes into sustainable agricultural food production systems. This will also imply minimizing the use of all the non-renewable inputs that are likely to cause threats to the environment, farmers, and ultimately consumers.

There are very few books that provide the necessary information on agricultural sustainability to farmers. This book covers materials that are extremely useful for the day-to-day operation of a farm or garden. It contains resources, techniques, and information that takes into account all the social, physical/environmental, and economic factors affecting agricultural practices.

It is primarily designed for all undergraduate and postgraduates studying sustainable crop and animal production. Extension workers, agricultural researchers, vegetable specialists, horticulturists, and other interested parties will also find the book to be useful. It can also be used to train small-scale farmers and gardeners in the agricultural sector. It is an adventure worth taking, since few things in life will improve general health and well-being as much as quality food products grown to their full potential.

For the convenience of readers, the whole book is divided into ten chapters. Chapter one deals with an overview on agricultural sustainability; chapter two uncovers factors influencing agricultural sustainability; chapter three deals with irrigation practices and agricultural sustainability; chapter four focuses on cultural farm practices that contribute toward agricultural sustainability; chapter five deals with sustainability of urban agriculture; chapter six focuses on dairy cattle keeping and agricultural sustainability; chapter seven interestingly uncovers beef-cattle farming and agricultural sustainability; chapter eight focuses on sustainability of poultry, rabbit, and fish farming; chapter nine is on arid and semi-arid areas potentials in agricultural sustainability; and chapter ten discusses the organic farming and agricultural sustainability.

All chapters contain a brief introductory section, and the following sections provide the detailed information. I sincerely believe and hope that the information presented in this book will be useful to you.

1 CHAPTER

AN OVERVIEW ON AGRICULTURAL SUSTAINABILITY

CONTENTS

1.1. Introduction..2
1.2. Definitions of Sustainability...2
1.3. Agricultural Sustainability Strategy...3
1.4. Past Studies on Agricultural Sustainability..4
1.5. Social Dimension Issues on Agricultural Sustainability.....................................6
1.6. Measurement of Agricultural Sustainability...9
1.7. Strategic Interventions Towards Agricultural Sustainability.............................13
1.8. Impacts of Biotechnology on Agricultural Sustainability..................................17

1.1. INTRODUCTION

This chapter presents basic information concerning agricultural sustainability. It serves as an introductory chapter to this book. Some of the areas discussed include: definitions of sustainability; agricultural sustainability strategy; past studies on agricultural sustainability; social dimension issues on agricultural sustainability; measurement of agricultural sustainability; strategic interventions towards agricultural sustainability; and impacts of biotechnology on agricultural sustainability.

1.2. DEFINITIONS OF SUSTAINABILITY

There are several definitions of sustainability directed towards the agriculture sector. The OECD (1995) defines sustainability as "a development strategy that manages all assets, natural resources, and human resources as well as financial assets, for increasing long-term wealth and well-being" According to the OECD, the value of natural resources should not be measured only taking into account their capacity to produce tangible goods, but also with the various non-marketed environmental benefits they provide. Trzyna & Osborn (1995) define sustainability with reference to development as "the complex of activities that can be expected to improve the human condition in such a manner that the improvement can be maintained" (Figure 1.1).

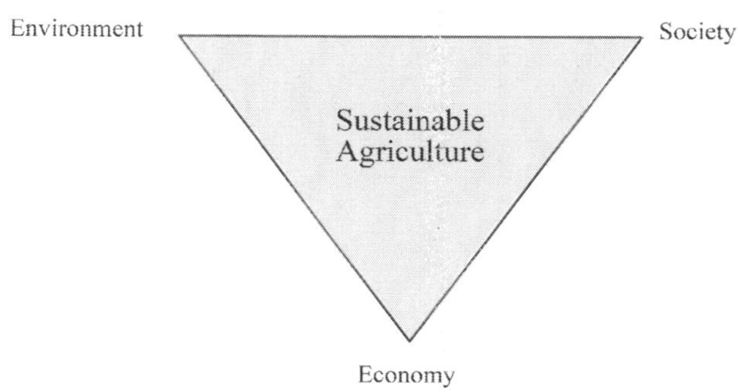

Figure 1.1: Illustrating key elements of agricultural sustainability (*Source:* https://www.nature.com/scitable/knowledge/library/sustainable-agriculture-23562787).

Sustainability is also thought of as the idea of maintaining some overall equilibrium between resource use and availability (Pierce, 1993). The American Society of Agronomy defines a sustainable system as "one that over the long-term, enhances environmental quality and the resource base on which agriculture depends; provides for basic human food and fiber needs; is economically viable and enhances the quality of life for farmers and society as a whole" (American Society of Agronomy, 1989). There is no conflict in the opinions expressed in these definitions. A synthesis of the definitions and an understanding of the concept indicate that a sustainable system is one that seeks to maximize present benefits while conserving resources (particularly natural resources) for future generations to use.

1.3. AGRICULTURAL SUSTAINABILITY STRATEGY

Agriculture is indispensable in both urban and rural development. It is an important component for development and sustainability. In developing countries, it has been referred to as the "backbone of our economy" as agriculture provides the second most important basic need, food (Henkel, 2015).

Agricultural practices need effective strategies in order to be successful and sustainable. As Trigo (1991) puts it, to succeed in achieving their goals, farmers need to formulate and implement strategies. These may be thought of as a pattern of purposes, policies, programs, actions, decisions and/or resource allocations that define the kind of agricultural venture that's performed, what it does, and why it does it. Johnson & Scholes (2002) define strategy as "the direction and scope of an agricultural enterprise over the long-term, which achieves an advantage for the enterprise through its configuration of resources within a changing environment to meet the needs of markets and fulfill stakeholder expectations."

The essence of an effective strategy thus becomes the purposeful management of change toward the achievement of competitive advantage in every agricultural venture in which it is engaged in. The strategies chosen by agricultural enterprise depend on factors such as leadership, resource availability and changes in the environment (Johnson & Scholes, 2002).

Aosa (1992) notes that the strategy being implemented should be realistic given the resources available. It is impossible to implement a strategy which imposes resource requirements that cannot be met by an agricultural enterprise. Therefore, it is important for any agricultural enterprise to pursue strategies that are capable of being sustained. Capacity for such sustainment is a measure of agricultural enterprise strategic capability. Its ability is provided by the resources that are at the disposal of the enterprise (Aosa,

1992). Those resources are deployed to the activities of the agricultural enterprise to create competencies.

1.4. PAST STUDIES ON AGRICULTURAL SUSTAINABILITY

The pioneering work in agricultural sustainability was carried out by Hayami (1969) for United States and later by Hayami & Ruttan (1971) for United States and Japan. The method employed in these pioneering studies and later studies in Korea by Ban (1979), Taiwan by Lee & Chen (1979), and Germany by Weber (1973), involved the agricultural production function approach and the productivity index approach alongside the "Induced Innovation Hypothesis."

The basic principle of "Induced Innovation Hypothesis" is the assumption that supply factors change, as reflected in the relative prices, which influence the direction and progress in agricultural development, affecting the adoption and eventual agricultural sustainability rate. In the course of agricultural sustainability, farmers are 'induced' to seek for methods that relieve the constraint imposed by the scarcest resource so as to increase their output. For instance, where land is very scarce, output is expected to grow through the use of yield-increasing technologies, while if labor is in short supply, mechanical techniques of a labor-saving type, tend to be adopted by farmers.

Using the "Induced Innovation Hypothesis," Hayami & Ruttan (1971) noted that in the United States where labor was scarce, relative to land, mechanical technologies in the way of farm machinery were adopted and led to high growth rates in output and in labor productivity, primarily due to increased land area per worker. In Japan where agricultural land was very scarce, the authors observed a progress in the use of biological technologies of a yield-increasing land saving pattern which accounted for growth in output per hectare.

Following Hayami & Ruttan (1971), other studies have been carried out in a number of countries, which have also indicated that factor substitution of technical inputs (Fertilizer for land, machinery for labor), for the scarce primary resources plays a big role in the growth of output and productivity. Yamada & Hayami (1979) also observed that improved crop varieties together with increased use of fertilizers and other current inputs that substitute agricultural land needs improved labor productivity as a result of increased output per hectare.

Similar observations were made for German agricultural growth by Weber (1973). He noted that similar to the "Induced Innovation Hypothesis" characterized in the United States and Japanese agricultural development, Germany sustained high rates of growth through a system of "Choice" of technology, whereby industrial as well as non-industrial products were used as substitutes for the scarce land resource. Labor was not scarce during that period (1880 to 1913) in Germany because of immigration of workers into Germany from the neighboring countries.

The central unifying feature, as far as agricultural sustainability is concerned, is the challenge faced by all countries. The necessity for sustaining rapid growth in output per capita at times characterized by high population growth, scarce land base and rising costs of yield-increasing technical inputs.

Yamada & Hayami (1979) emphasized the role of biological innovations that improve a plant's capacity to respond to fertilizer use, in their explanation for high land productivity in Japan. In the Philippines, Crisostomo & Barker (1979) did not observe any direct relationship between changes in the level of non-farm current inputs and land productivity. Instead, they observed that growth in land productivity was mainly as a result of changes in yields of the major crops, whereby food crops rather than export crops were the chief contributors. Thus, the use of technical inputs alone resulted in only a modest growth in output, while changes in the product mix and product performance due to breeding and selection had a significant effect on the growth of production in the Philippines.

The authors of the Taiwan study noted remarkable advancement in agricultural technology as the basis for the high output growth (4.2%) and land productivity gains (3.9%). Similar to the Hayami-Ruttan model, the use of new techniques for the scarce land resource was responsible for the observed growth rates in Taiwan. The authors noted that the high growth rates were a result of strategies involving the use of improved crop varieties, pest and disease control, improved cultural methods, and irrigation facilities coupled with the intensive use of fertilizers.

Similar remarkable achievements in technology adoption were observed in Korean agriculture by Ban (1979) who noted that the level of education and skill among farmers was a contributing factor to the high productivity of the farming community in Korea.

1.5. SOCIAL DIMENSION ISSUES ON AGRICULTURAL SUSTAINABILITY

1.5.1. Human Capacity and Agricultural Sustainability

This refers to human beings using their capability skills and knowledge in order to achieve higher and more sustainable levels of production. Knowledge, skills and developed values are enhanced capabilities of the people for a constructive organization in the agricultural sector. To ensure food security and appropriate nutrition of the ever-increasing global population, countries such as the United States of America and the United Kingdom have adopted new agricultural farming strategies.

According to Agricultural Extension Society of Nigeria (2015), the application of knowledge and skills is used as efforts to improve the quality of life, workers' rights, democracy, and sustain cultural traditions (gender equity, adult literacy). For instance, urban agricultural in most places around the world has been enriched by the skills and technologies of immigrant populations. USA, United Kingdom, and the Netherlands have provided training courses to enhance food production in sub-Saharan countries such as South Africa, Tanzania and Kenya.

According to the World Bank (2010), in the United States of America, skills and education at all levels lead to productive growth and better living standards. This shows that most farmers with basic education are bound to be productive. The main reason for stunted economies in most of the African countries are due to low skilled human capital. The quality of human capital is dictated by the quality and quantity of education, health and nutrition available to the people as stated by the World Bank (2010). Therefore, investment in human capital is important because it´s bound to improve the quality of labor and productivity in agricultural sector. Furthermore, market chain development should be stimulated in most countries. This can be done through differentiation and diversification when adopting a market strategy, processing and distribution system.

1.5.2. Social Capital and Agricultural Sustainability

According to Putnam Zhen & Zoebisch (2006), social capital can be defined as "features of social organization, associated norms and values, which create externalities for the community as a whole." Through social capital, like-minded people are able to come together and reinforce homogeneity.

Most farmers, especially in developed countries are able to evolve complex alliances in relation to local economies that shape livelihoods in their society. This starts with labor on farms and social networks on marketing of the agricultural products from the farms.

In United States of America for instance, the community-gardening organization provided job training. The pace, quality, and magnitude of most agricultural activities are influenced by partnerships formed by:

- Individuals
- Private organizations
- Public groups and governments departments

Social capital in the agricultural sector also contributes to community's well-being and solidarity. Neighborhoods within farm systems have higher levels of social interaction and enforced security. This contributes to the success of agricultural activities by the relations of laborers. Therefore, farmers can work within groups to:

- Reduce costs of production
- Develop skills and
- Enhance knowledge for agricultural sustainability

Studies have demonstrated that strong areas of social input are normally voluntary and a strong existence of linking social capital in production, organization of resources and consumption in agriculture for different households.

It has been shown that social capital, especially in developing countries, is helping the weakest members of poorer communities. Disaggregating social capital reveals diversity among membership, focus and a positive strong influence on the per capita expenditure of households. Therefore, there is the need to strengthen it through active facilitation of relationships, providing people (farmers) with basic agricultural information, material resources, technical support and solidarity as they are crucial for effective agriculture practices (Ragasa, Catherine, & Mariam, 2016).

1.5.3. Access to Resources and Agriculture Sustainability

Global population growth has been very rapid. This has led to competition over resources such as land, water and even human resource required for agriculture. According to World Bank (2006) report: lack of access to capital and land are a barrier to agricultural sustainability. This is common in areas where the land faces challenges of urban sprawl. Lack of access to essential agricultural land normally causes farmers to be reluctant to invest in their plots.

In sub-Saharan Africa most poor are found in the informal settlements and they normally have to cope with several challenges such as:

- Lack of farmlands
- Access to water
- Land tenure issues

The economic and social research council in the United Kingdom indicated that access to land for agriculture is a challenge. These changes affect the nature of agricultural activities in many places.

1.5.4. Participation and Agricultural Sustainability

There are several initiatives that local farmers can undertake as a way to show the local government their capabilities. The work could extend to education, welfare, agricultural projects and issues of governance. This could be through:

- Interest groups/pressure groups/lobbies
- Promotion groups
- Volunteerism
- Social groupings and
- Social activism

High levels of civil capital help in the creation of effective agricultural institutions, where the participation of people in social activities is essential. The National Research Council (2010) indicates that up to 70% of farmers in developing countries produce entirely for own consumption. Similar studies state that planning and governance in agricultural sector are shaped by the congruence of interest groups with conflicting interest competing for resources. Therefore, there is the need to build partnerships for the poor in order to make access to land, infrastructure, and finance affordable. Farmers should also work in networks in order to secure their interests and reduce the cost of production and consumption. Additionally, farmers should engage in a wide range of agricultural activities. This will ensure that agricultural sector remains sustainable on the long-term.

1.5.5. Poverty and Agricultural Sustainability

Most farming activities, especially in developing countries are undertaken by the vulnerable poor who are approximately 50% of the rural and urban population living below the absolute poverty line (Zhen & Zoebisch, 2006). Poverty makes the poor to engage in poor farming activities.

- Poverty is characterized by:
- Low incomes
- Lack of access to income assets and
- Poor environmental condition

The poor have higher per capita food production and their units of production are usually in competition with market forces. As such they suffer from food insecurity, access and adequacy. The poor possess assets such as labor, human capital, household relations, social capital and these complex assets affect their vulnerability to food access and quality. Sustainable food production is therefore a crisis strategy for survival of the poorer segment of the population (Wright, 2012).

The unique nature of poverty manifests in the form of several vulnerable groups: 40% of women headed-households, 5% of children headed households and 55% of non-committed labor force. Therefore, there is the need for organizing agricultural resources in order to counter various challenges caused by rapid and dynamic population growth rate.

1.6. MEASUREMENT OF AGRICULTURAL SUSTAINABILITY

There are a number of approaches applied towards the measurement of sustainability. These approaches are based on the definition and understanding of what constitutes sustainability. One such approach views agricultural sustainability as food self-sufficiency (Trzyna & Osborn, 1995). In this approach, estimating the future demand for food is the essential first step in achieving sustainability.

Once these estimates are available the next step is to assess supply capacities of the production system so as to estimate the supply of economically useful agricultural resources, and their productivities. Thinking economically about agricultural resources in the future leads to the conclusion that the supply of food resources can always be increased at some cost. Physical limits of resources almost never constitute an issue of particular products or sectors of an economy as long as society is willing to incur the economic and/ or environmental costs of recruiting the resources from other potential uses.

There is another group of people who focus their understanding of agricultural sustainability primarily on ecologic insights rather than on economic insights. These people see environmental costs rather differently. The ecological view of sustainability arises from the belief that nature, in the long run, imposes limits on humankind's collective capacity to provide food for the world. This group, therefore, views sustainability in terms of food self-sufficiency just like the first group. The limits imposed are physical in nature because they are imposed by a finite supply of natural resources and by the limited capacity of our physical environment to absorb waste. In this view, another measure of agricultural sustainability, therefore, is the sustained yield capacity of renewable agricultural resources. To maintain the long-run productive capacity of land for example, farmers need to maintain the depth of topsoil as well as a favorable root zone for plants, and they need to ensure other physical, chemical and biological conditions for the growth of plants (Sutherland, Wilson, & Zagata, 2014).

Some approaches for measuring sustainability are limited to a specific type of quantifiable variables. In others both control and state variables are quantified. State variables are descriptors of the quality of the environment or of specific resources while control variables are those that directly influence the level of state variables. For example, the control variable "tillage practice" influences the state variable "soil depth remaining after erosion." There is usually a cause-effect relationship between state and control variables. When only state variables are measured, considerable doubt can remain upon the causes of observed changes.

The measurement of sustainability may be quantitative or non-quantitative. Proponents of non-quantitative approaches seem to reject the notion that sustainability can or should be measured. Trzyna & Osborn (1995) argue that quantification tends to distort the research process, inducing researchers to choose quantifiable (but less relevant) variables at the expense of other non-quantifiable (but conceptually more important) ones. They are especially skeptical of numerical modeling of biological systems, arguing that the internal consistency of these models does not compensate for their lack of realism. However, it is never possible to deal with any problem (not just sustainability problems) in all its real-world complexity. According to McCall & Kaplan (1985), scientists "have to simplify to survive." In addition, the experience of farming systems research suggests that it is often possible to quantify and model complex biological systems without unacceptable loss of realism.

While most proponents of sustainable agriculture would probably not agree that measuring sustainability is utterly impossible and that trying to measure it is a bad idea, many however, would be content with "directional measurements." A directional measurement is one that measures only the direction of change in the sustainability of a system, not the magnitude of that change. Directional measurements are most attractive when it is felt that a proportional relationship exists between control and state variables. The assumption is that the sustainability of an agro-ecosystem is changed in rough proportion to those practices felt to most strongly influence the system's future productivity (and/or its ability to deal with stresses and perturbations). For example, an agro-ecosystem suffering from gradually declining levels of soil nutrients is thought to become more sustainable in rough proportion to the amount of nutrients that through appropriate interventions, can be generated or recycled within the system or applied from external sources (vanLoon & Hugar, 2005).

1.6.1. Agricultural Projects Sustainability Monitoring and Evaluation

The concept of monitoring and evaluation refers to a combination of various activities, including:

- Planning
- Capacity building
- Surveillance and
- Information utilization

These activities may viably contribute to decision-making, learning and agricultural sustainability. This means that undertaking monitoring and evaluation activities, especially in agricultural projects, can be used to enhance realization of sustainable agricultural practices.

Availability of trained monitoring and evaluation personnel is one of the major limitations in most countries around the world (vanLoon & Hugar, 2005). Other existing challenges to agricultural sustainability monitoring and evaluation include:

- Lack of mechanisms to assess the skill gaps among personnel functioning in the Monitoring and Evaluation area.

- Incompetence among organizations and personnel.
- Scarcity of monitoring and evaluation staffs.
- Lack of quality evaluations.
- Lack of meaningful authentication of monitored data leading to reliance on survey-based.
- Poor data analysis.

In developing countries, the main challenges for agricultural sustainability monitoring and evaluation is the promotion of transparency and surveillance which is directly at the heart of challenging political hegemonies contrary to the advocacy of the theory of social change regarding inclusivity. Freedom to present findings in a public domain may be censored or fully prohibited (Trigo, 1991). This tends to weaken surveillance, a key ingredient of monitoring and evaluation. Such practices definitely affect the relationship between monitoring and evaluation of agriculture practices towards attaining sustainability objectives.

Studies have shown that most agricultural projects funders impose sustainability and capacity building as fundamentally integrated into the project planning and design, to ensure that when funding is withdrawn, the project's activities and positive impacts will continue. In agricultural projects, sustainability is the ability of a project to maintain its operations, services and benefits in the society (Henkel, 2015). The issue of sustainability is considered in light of changing social economic political contexts. Historically, donors have treated development and sustainability separately, a practice that has aggregated the concept of poverty, especially given the failure of interventions calculated to alleviate poverty.

Therefore, understanding sustainability has tended to be biased towards natural resource management. However, sustainability should focus also on such issues as planning, capacity development and technology transfer (Trzyna & Osborn, 1995). This view is in line with theory of change. Efforts aimed at facilitating sustainability should focus on the people at the grassroots level. The concern that arises is on how ready the state and local structures are towards empowering vulnerable groups, even when sponsors withdraw.

Agricultural sustainability monitoring and evaluation systems to be useful, they should be driven by demand and not supply forces. Making use of monitoring and evaluation results and information is a major determinant of sustainability. It is also a product of good planning, project implementation that is based on availability of capacity and informed decisions based on sound and relevant information (FAO, 1992).

It has also been noted that monitoring and evaluation provides useful information that can be fed back into an agricultural project. This information can be used to:

- Improve policy analysis and development.
- Aids in project activities.
- Help in managerial activities.
- Enhance transparency, surveillance and agricultural project sustainability.

If the monitoring and evaluation officials and beneficiaries are deficient in capacity, this will ultimately impact on sustainability of agricultural projects and hinder the empowerment process that should be intertwined with development interventions in tandem with the theory of social change and utilitarianism.

1.7. STRATEGIC INTERVENTIONS TOWARDS AGRICULTURAL SUSTAINABILITY

The success and sustainability of agricultural sector highly depend on proper planning, management, and priorities setting of all agricultural activities. This in other terms means that there is a lot that needs to be done in order to realize full potential of agricultural sector. Some of the priorities that can be set in order to achieve full sustainability in agricultural sector are discussed below.

The main intervention toward agricultural sustainability is to train farmers on proper farming activities and land consolidation. Land consolidation uses will make it easier towards agricultural mechanization phases because of economy scale. The use of agricultural input is very important in improving agricultural productivity and sustainability which is part of a modern agriculture (Hayami & Ruttan, 1971).

Another intervention towards agricultural sustainability is to sensitize and train farmers on use of agricultural input and on basic mechanization technics. To achieve this action, the government should increase training of extension agents.

Develop a system where private sector plays a big role in providing agricultural input. In the case of selected agricultural inputs, a partnership between the stakeholders involved should be initiated in some particularly areas. In terms of livestock development, the increase of livestock should be accompanied by the improvement of exotic breeds that will help in maximizing production of essential livestock products such as milk, meat, eggs, wool, skin and hide (Abington, 1992).

1.7.1. Agricultural Land Use

Sustainable agriculture land use is not currently being optimized because of the scattered nature of settlements in some rural areas. While this is a medium to long-term process, these problems create immediate shortfalls in productive activities, particularly for agriculture production. The following priority actions should be taken to sustain agricultural land use.

Land consolidation: Land use consolidation is one of the policies that enable an efficient use of agricultural land in terms of agricultural mechanization. Sensitize farmers on the benefits of land consolidation should be the main action.

People's rights to land: Effective implementation of agricultural land use should be based on strong protection of people's rights on land. According to Lichtfouse, Navarrete, & Debaeke (2009), the community needs to be involved in the planning and decision-making process, not simply on the implementation plans. In other words, this should allow rural farmers to lead in the planning and economic development of their agricultural land. This is a major way to ensure sustainable growth in agricultural sector.

Rural infrastructure: Scattered rural living means that farmers in rural areas are unable to connect to vital agricultural services. Connecting communities to modern infrastructure will be much easier and cost-effective in grouped settlements. Returns to scale are paramount here. The more dense rural settlements can incentivize agricultural productivity, as it will be more feasible for rural farmers to access modern infrastructures. For example, creating roads to transport their farm produce to markets. However, it is important to ensure that infrastructure and services reach these farmers, old or new (Trigo, 1991).

1.7.2. Access to Finance in Rural Area

Use of financial services by the rural population of formal, semi-formal and informal types is important for agricultural sustainability. There are several actions that can be taken in order to solve the issue of financial constraints in agriculture.

Promoting financial literacy: Financial literacy promotion is needed to ensure that farmers understand the products and the advantage of using financial services in their daily agricultural practices.

Encouraging informal financial products: This will allow small-scale farmers to understand the importance of savings. Additionally, these financial products will enable more farmers in the rural areas to enhance their financial literacy, especially towards cash management.

Strengthening financial Saccos: It is highly important that governments improve local financial Saccos and financial branch towards sustainability. While strengthening the Saccos, a diverse range of products for rural households should be promoted such as branchless banking products, remittances, transfers, payment and insurance.

1.7.3. Access to Agricultural Information

The full realization of the potential of agricultural sector has been hindered by a number of factors including lack of access to production, value addition and marketing information. Most farmers especially in developing countries do not have access to the right information. This information includes:

- Chemical application rates.
- Managing diseases.
- Where to access certified seed.
- The best varieties fit for specific location.
- Housing and managing livestock among others.

Therefore, there is the need to setup communication strategies in order to achieve an effective use and sharing of the right agricultural information. Additionally, different scholars have suggested the need for participatory approaches and strengthened collaboration mechanisms in the delivery of agricultural information (Figure 1.2).

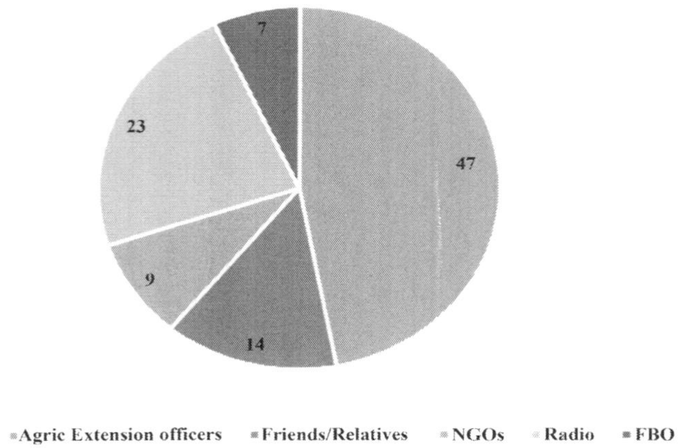

Figure 1.2: Source of agricultural information (*Source:* https://www.researchgate.net/figure/Sources-of-farmers-Agricultural-information-Source-Field-Survey-2016_fig2_317434432).

With the necessary information, farmers in rural areas can produce enough for consumption and sale, hence helping a region/country realize agricultural potential. The use of information in agriculture creates a digital divide in the sector and early adopters normally benefit a lot. Information in agricultural sector also enables farmers in becoming good managers of their farms and thus improves productivity. Most farmers in rural areas are wallowing in poverty despite having plenty of resources. Therefore, decision-making on who does what in the agriculture sector is important and should form a strong base in the planning for the sector (Ruben, Pender, & Kuyvenhoven, 2007).

Access to information can be enhanced through on-farm demonstrations, shows and exhibitions. This can further be strengthened through collaboration mechanisms by creating a strong link between the ministry/department responsible for agricultural support and service organizations involved in the transfer of necessary knowledge (Trigo, 1991).

However, there are key issues affecting the agriculture sector in terms of information access in many countries. Key to them is inadequate demand-driven research coupled with ineffective extension and transfer system of research findings. Therefore, strengthened information infrastructure system to transfer findings is a sure way of creating a strong linkage and hence more agricultural productivity (Sutherland, Wilson, & Zagata, 2014).

Therefore, all relevant stakeholders should be identified, analyzed and their involvement and responsibility levels established.

1.8. IMPACTS OF BIOTECHNOLOGY ON AGRICULTURAL SUSTAINABILITY

The advancement in agricultural biotechnology has enabled scientist to come up with a new agricultural crop varieties, which are seen as a big step in solving food crisis in the world. According to FAO (1992), agricultural biotechnology refers to "the application of scientific research tools to understand and manipulate the genetic makeup of organisms for use in agriculture." Biotechnology is actually one latest scientific tools that have been adopted which helps in developing new crop varieties that help in improving agricultural productivity by reducing production costs and enhancing yields.

Some of these biotechnological tools include:

- Genetic engineering,
- Embryo rescue and
- Tissue culture.

For instance, NERICA rice is one of the new rice varieties that have been developed by the use of embryo rescue which involved crossing two species of rice ("the African rice *Oryza glaberrima* with the Asian rice *Oryza sativa*"). Coming from different species the resultant embryo would not have survived naturally, hence the need to 'rescue' the embryo by culturing it in a suitable medium that allowed its development to a viable plant. The idea behind NERICA rice development was actually to create a new type of rice which could do well under various climatic conditions such as inadequate farm inputs, frequent droughts and numerous diseases and pest. Some of the desirable traits of this new type of rice include:

- Better response to fertilizer,
- Tolerance to low moisture stress and
- Higher protein content.

1.8.1. Agricultural Research

Most of the farmers in the world, especially in developing regions are poor. They practice smallholder farming in unfavorable agricultural zones such as arid and semi-arid areas. This calls for more support to alleviate agriculture in these areas. This is so because farming is the main source of livelihood for the majority of the population and supporting it will mean making agriculture sustainable for the majority of the citizens. According to the World Bank (2008) report, it has been noted that growth in agriculture is vital for stimulating growth in other sectors of the economy.

According to Trigo (1991), one way of increasing growth in agriculture is through adoption of superior technologies. It is therefore significant for those countries practicing agriculture for sustainability to invest in research which aims at delivering superior agricultural technologies. This will ensure enhanced income generation and reduce poverty levels.

1.8.2. Innovation of Agricultural Biotechnology in Small-Scale Farming

Recent agricultural biotechnology has been advocated as a tool to make crop research benefit poor farmers in marginal areas. As stated before, small-scale farmers have unique challenges and thus need really innovative technologies to overcome them. According to FAO (2004), the term innovative technology can be described as a marketable good which allows farmers to surmount an agricultural problem. However, before adopting any technology in the agricultural sector, it is important to always assess whether the technology is innovative enough to handle most of the unique constraints that small-scale farmers face.

There are several benefits of biotechnology crops in the agricultural sector. Some of these benefits include:

- Biotechnology crops increase yield.
- Biotechnology innovation reduce the overall costs of pest control.
- The total costs for weed control is also minimized through biotechnology application.
- Most genetically modified seeds normally cost more than the conventional seeds.

Considering the fact that most biotechnology innovations in agriculture have the ability to impact on both the cost and output elements of production, there is the need for all farmers and policymakers to understand the net effect of biotechnology adoption. It is crucial for them to determine whether biotechnology is innovative enough to overcome production challenges and deliver higher net returns to small-scale farmers.

Bhullar (2012) showed that modern agricultural biotechnologies such as the use of genetically modified crops have high potential to offer substantial benefits, especially in developing countries where agricultural crop production is faced with a lot of challenges. Such potential has been observed in China and India where most of the biotechnology cotton production is by small-scale farmers. Impact studies in these countries have revealed that farmers benefit enormously from adopting the biotechnology innovation.

Where nationwide benefits of biotechnology adoption are to be evaluated, the extent of benefits may sometimes be affected by the importance of the crop to the farmers and the level of adoption. This was observed by Lichtfouse (2014) that reported welfare gains of between 0.8 and 16 million Euros in Hungary among three crops namely transgenic maize, sugar beet and oil seed rape. Adoption of biotechnology in a crop that is more important in terms of size of production was found to give more welfare gains to the nation than its counterpart which is grown on a lesser scale.

In addition to the direct economic benefits, there are also environmental benefits that biotechnology has potential to deliver. For instance, studies have shown that the use of genetically modified crops normally have a significant impact on the reduction of total costs incurred in the control of weeds in some crops such as soybeans and corn.

There are, however, other cases where the same biotechnology application has shown different results under small and large-scale farming conditions. Taking as an example the adoption of herbicide-tolerant soybeans, researchers have not found statistically significant differences in profits between genetically modified and conventional soybean farmers in the US. It was, however, observed that adopting the genetically modified soybean increased farmers' profitability by an average of 19 Euros per hectare in Argentina, representing an increase of 8.5% over the gross margin obtained by conventional soybean farmers. It was also demonstrated that the increase in gross margin was higher for the group of smaller farmers than for larger farms in Argentina (Trigo, 1991). This may be due to higher efficiency among smallholder farmers.

Notwithstanding the potential benefits of biotechnology innovations in crops production, there are still concerns with respect to the sustainability benefits over time. Some of these concerns include:

- Researchers and farmers fear that the population of most pests after certain period of time may become resistant to the biotechnology innovations.

- There are fears that most of the minor pests that are not controlled through biotechnology innovations may sooner or later turn to become prime pests.
- Researchers have fears that most dominant market structures might increasingly prevail due to commercialization of the genetically modified crops.

Concerns have been raised regarding distribution of benefits arising from adoption of biotechnology. The question of distributional equity is especially important nowadays since wealth distribution is becoming an important measure of economic development in any society. Many opponents of biotechnology have argued that multinational seed companies will take all, if not most of the benefits arising from biotechnology. They argue that biotechnology companies will use their monopolistic powers to price products in a way that makes the cost of crop production high. Thereby lowering farm profits as well as raising food prices. Such a scenario would undermine both farmers and consumers economically. But empirical evidence has revealed that 50–70% of these benefits go to farmers and consumers while the rest go to the seed companies (Macnaghten & Ripalda, 2015).

As such, most of the studies on the impact of biotechnology indicate more benefits than cost. There are, however, some concerns that are difficult to measure, for example, some religious groups have raised questions regarding the ethical correctness in modifying God created organisms.

1.8.3. Impact of Biotechnology Among Small-Scale Farmers

A particular biotechnology innovation may have different effects on different communities. These effects are determined by the uniqueness of the communities based on the socio-economic conditions they are facing (vanLoon & Hugar, 2005). Studies have also shown that most effects of biotechnology innovation are highly dependent on the methods used. Whereas some studies have focused on the environmental impacts (both negative and positive), some have restricted themselves to the measurable monetary benefits. Information on the effects of biotechnology innovations can be used by policymakers in developing nationwide strategies. However, for farmers, the most important unit of analysis is his or her farm business (Aosa, 1992). Given the increasing demand for bottom-up approaches to technology development and adoption, the economic surplus model, which does not give farmers specific information, may not be the most appropriate.

Biotechnology may have both marketable and non-marketable impacts. When evaluating the economic effects of biotechnology in agriculture, there are some fundamental aspects that should be tested. These include:

- Changes in yield,
- Amounts and costs of inputs and
- Profits.

Change in incomes related to application of biotechnology innovation derives from cost reduction, yield increase, and price shifts. Small-scale farmers driven by profit the information regarding inputs costs, yield and profit dynamics is essential (vanLoon & Hugar, 2005).

Most of literature and government strategic plans have consistently identified resource limitations as a major impediment in small-scale farming. Resource limitation has often been worsened by the fact that most small-scale farmers engage in multiple farming enterprises simultaneously. This has sometimes made farmers allocate too little resources to each enterprise, resulting in the loss of economies of scale and so suffering loses. It is therefore crucial that impact assessment tools are tailored to capture this twin condition of resource limitation and multiplicity of farming enterprises. In most cases, farmers and policymakers need to know the most profitable way to allocate the scarce resources.

There are several conditions that must be considered when evaluating the effects of biotechnology innovation in agricultural sector. Some of these conditions include:

- The agro-climatic requirements (soil type, water requirements, temperature, humidity and pests and disease levels),
- Input requirements,
- Market potential
- Government policies & regulatory framework and
- The available farm level resources (land, labor capital).

After considering the prevailing conditions, a new technology is then assessed against the existing practices to determine whether it is actually superior. In case there is potential for positive change in farm income, farmers respond to the availability of such superior technology by allocating more resources to it (Dillard, Dujon, & King, 2008). This means changes both in the total farm income and the relative contribution of the various enterprises to the total farm income (Figure 1.3).

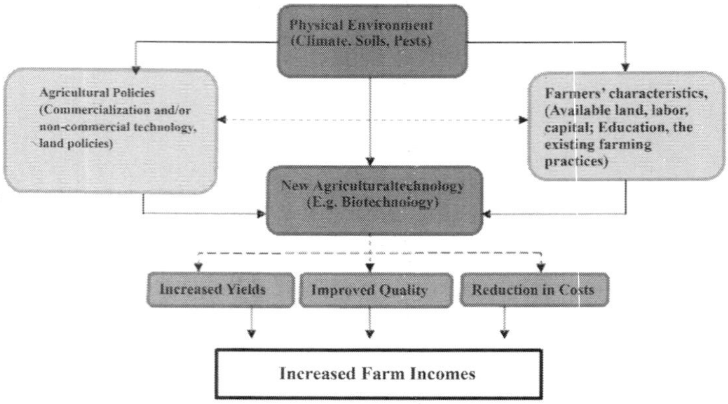

Figure 1.3: Illustrating the interaction between biotechnology innovation, prevailing conditions and effects on agriculture practices (*Source:* https://www.ecologyandsociety.org/vol4/iss1/art13/).

1.8.4. Challenges in Agricultural Technologies

Despite the many benefits that come with agricultural researches and adoption of new technologies in agricultural sectors, most poor farmers, especially in developing countries, have not yet benefited from these benefits. This is because most of these farmers are practicing their agriculture activity in harsh environments which are usually characterized by:

- Unreliable rainfall.
- Poor soils.
- Heterogeneous topography.
- Lack or poor infrastructural facilities such as bad roads.
- Limited credit facilities.
- Lack of irrigation facilities.

Lack of adequate information on the available superior technologies is another reason why poor farmers usually miss out on the benefits of agricultural technologies (Sutherland, Wilson, & Zagata, 2014). This is because new and superior technologies are comparatively more expensive than the existing less productive technologies. Research has also revealed that the lack of consistent produce can also be a hindrance to adoption of new technology in agricultural sector.

From this discussion, it can be concluded that despite the fact that technological change is useful to farmers, there are some major limitations which should be addressed if small-scale farmers are to benefit from such technological advancements. The superiority of each technology should, therefore, be assessed against the prevailing farm level opportunities and constraints. In other words, all the socio-economic conditions surrounding agricultural sector should be taken into account.

2
CHAPTER

FACTOR INFLUENCING AGRICULTURAL SUSTAINABILITY

CONTENTS

2.1. Introduction 26

2.2. Resources 27

2.3. Climatic Factors 27

2.4. Financial Factors 28

2.5. Land Fragmentation 30

2.6. Technological Factors 33

2.7. Information Access 36

2.8. Social-Cultural Factors 37

2.9. Region Factors 38

2.10. Cropping Pattern 39

2.11. Storage Facilities 40

2.12. Agricultural Cooperatives 41

2.13. Policy Formulation 42

2.14. Marketing And Pricing 42

2.15. Agricultural Extension Services 43

2.1. INTRODUCTION

Sustainability implies the long-term continuation of an enterprise. Agricultural sustainability needs to be understood in terms of several elements or components without exclusive focus on any single element and its important to analyze these components based on their ability to be self-sufficient.

As Acton (2013) put it, the concept of sustainability has financial and institutional dimension. There are broad issues of resource capability which are relevant to an agricultural enterprise as a whole and these are largely concerned with the overall balance of resources. Acton (2013) argue further that agricultural sustainability is more than financial since the impact of the work has to be sustained in the long run. Agricultural sustainability thus cuts across several aspects of an agricultural enterprise and essentially entails resource possession or control, utilization and configuration towards agricultural goals. These resources include physical, financial, human, intellectual, marketing, research and development, and information. According to Johnson & Scholes (2002), resources in agriculture include those that are owned by the enterprise and those that can be accessed to support its strategies. This is because some strategically important resources may be outside the ownership of an agricultural enterprise, such as its network of contacts or customers.

Bearing this in mind, sustainability is difficult to quantify as it requires a long-term view of the agricultural development process and implies the long-term continuation of an agricultural enterprise. Agricultural sustainability can be seen, especially among fully functioning agricultural enterprises, as having diversified resource base and partnership relationships with national and international networks.

There are several factors discussed in this chapter which are relevant to agricultural sustainability in one way or another. These include: Resources, Climatic factors, Financial factors, Land Fragmentation, Technological factors, Information access, Social-cultural factors, Region factors, Cropping Pattern, Storage Facilities, Agricultural Cooperatives, Policy Formulation, Marketing and Pricing, and Agricultural Extension Services.

2.2. RESOURCES

In the modern age, the most highlighted sustainability issue concerns the inadequacy of resources. According to (OECD, 2011), the population of the world has grown tremendously over the last few years. A growing population signifies that the available resources will become strained. Many governments and other stakeholders have downplayed the seriousness of the issue by assuring the citizens of better lives in the years to come. Already, some private agricultural entities have started to research on alternative ways of acquiring the scarce raw materials used to develop various agricultural products.

The absence of natural resources will mostly affect private agricultural entities and ultimately the world governments. Large governmental agricultural entities will strive towards blocking the activities related to resource scarcity. The small farms will likely collapse first when the raw materials become so much strained. The scarcity will eventually impact on the lives of the people around the world causing civil and political unrest. The lack of necessities weakens the influence that the government has on people by building feelings of resentment that culminate into anarchy movements (Hayami & Ruttan, 1971). Just like it happens in the case of corporations, the weaker developing countries will fall first when the shortage of natural resources increases.

2.3. CLIMATIC FACTORS

Climate change is another sustainability issue that has raised heated debates around the world. Climate change involves the warming of the earth due to increased carbon dioxide and other gaseous emissions that create a greenhouse effect. The concentration of such gasses in the atmosphere creates a blanket that traps heat within the earth. The rising temperatures lead to the melting of polar ice caps which triggers and results in rising ocean levels (Stewart & Coclanis, 2011). The lifting of the sea levels submerges many parts of the continents where people live, and the available land capital reduces (Figure 2.1).

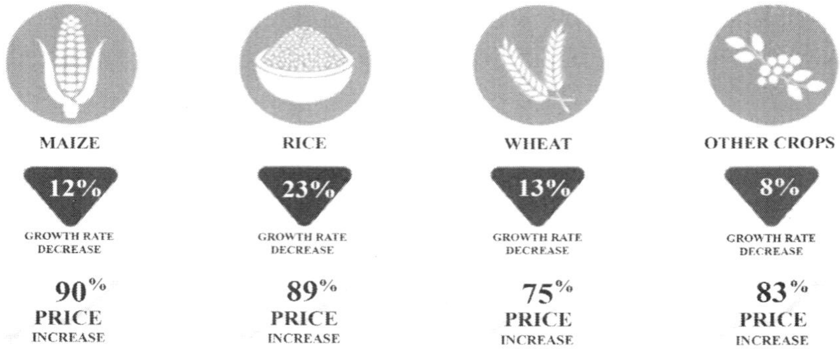

Figure 2.1: Illustrating impacts of climate change on agricultural sustainability (it is estimated that 10% of major crops will experience a stagnant or reduced growth rate by 2030 due to climate change) (*Source:* https://farmingfirst.org/sdg-toolkit#section_1).

Climate change also results in adverse weather patterns. The introduction of carbon dioxide in the air changes the seasons, and therefore agricultural companies suffer from the unpredictable seasons. Cultivation becomes more complicated as crops dry before their time. The rising ocean temperatures increase the number of hurricanes experienced in any location. Scientists have attributed the recent surge of tornadoes in the United States to climate change. Hurricanes and storms cause significant destruction to property and other business assets (National Research Council, 1993). Businesses that lack insurance cover collapse immediately and others that have indemnity take long to regain their momentum.

2.4. FINANCIAL FACTORS

According to Acton (2013), when one talks of agricultural sustainability, financial sustainability comes to mind for two reasons, one is that this would secure their autonomy and another is that outside funding often dries up. He adds that sustainability cannot be seen on the basis of financial sustainability alone, it has to also include social & cultural sustainability, political sustainability and technological sustainability.

Other than financial self-sustainability, aspects of agricultural management structure are also important in ensuring agricultural processes provide services in the long-term, the agricultural structure may be the most significant design element in relation to long-term sustainability, Acton (2013) further observes that considerations of sustainability in financial terms need to be made from the onset, but even when the program is already underway, prospects for financial and organizational sustainability should be addressed, and it should be done within the national regulatory framework, as it may be necessary to undertake advocacy work to change existing government policies.

There have been several studies undertaken where most concentrate on the financing pressures and little on the concept of sustainability with respect to agricultural practices. Sustainability being more than financial, most literature looking on agricultural practices thus falls short on the concept and looks only at one aspect.

Sustainability strategies in agricultural financing includes concerns for the quantity or level of funding, timeliness or stability in funding, financial accountability and the use of funds for issues or activities that are a priority for stakeholders and customers of an agricultural enterprise (Bhullar, 2012). He adds that developing and managing plans, mechanisms and sources for such strategies involves processes of coalition development, strategic planning and institution innovation.

The financing of agricultural services, and especially of research and extension, attracts much attention in the development community. Agricultural research and extension in many countries are looked at together and carried out by similar organizations. Public funding for agricultural services is getting scarcer than ever and governments as well as donors are engaging in new forms of funding. On the recipient side of the funding system, more and more advisors and their partner institutions are faced with the need to raise money independent of conventional budget allocations.

In funding, the first issue concerns public and private roles in development managing funding levels for public services is not just a question of tax income but of principle.

Governments need to review their portfolio of services and decide whether to discontinue funding in some cases or have recipients pay in others. Secondly, rational funding has to use the economic principle, i.e., make the best use of scarce resources. Efficient allocation and control of funds is an important concern in any funding arrangement (Zhen & Zoebisch, 2006) (Figure 2.2).

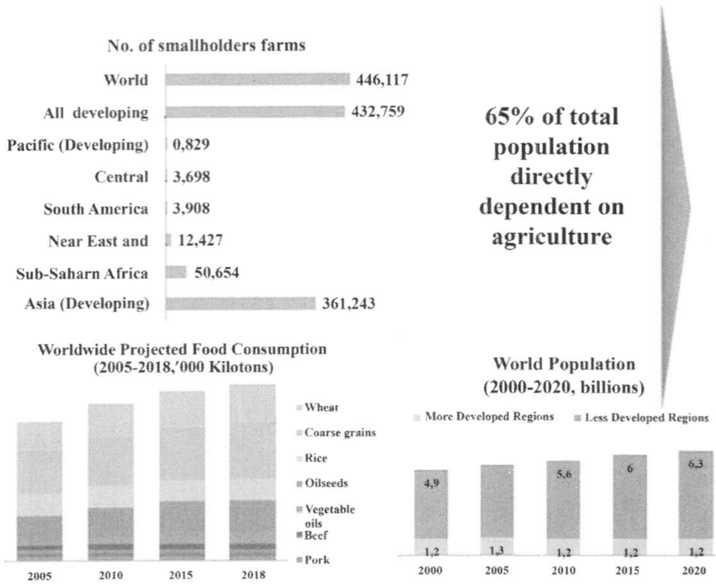

Figure 2.2: Sustainable agriculture is a key for economic development; the sector requires more financial assistance (capital and capital investment) in order to meet all requirements of population with higher purchasing power (*Source:* http://slideplayer.com/slide/8033460/).

Mirroring the change on the side of the funding sources, a third issue focuses on the increasing pressure resting with service providers to seek new funding sources. This requires additional qualifications and a new way of looking at stakeholders, which should ensure an enhanced client and development orientation of service institutions. Reversing the direction of decision-making or reversing the flow of funds is the fourth topic of great interest. In the interest of sustainable development, there is the need to get away from a situation in which the money, the plans for using it and the services themselves are all dominated by the supply side (Zhen & Zoebisch, 2006).

2.5. LAND FRAGMENTATION

Land is man's most valuable resource, in which his existence and progress depend. According to Ruben, Pender, & Kuyvenhoven (2007), "it is what people divide, occupy and assert rights to, socially and politically in order to exploit it economically." Land can be acquired through buying, inheritance, renting or as a gift. The new demands for land and space continues to create great competition for the remaining stock of this resource. Land fragmentation is a common phenomenon in different parts of the world, where these particular patterns differ from country to country. A few countries practice fragmentation of land use, where a household operates on separate parcels of land on a single farm, but the majority (especially from the developing economies), practice fragmentation ownership where a single farm has been subdivided into plots, each with its rightful ownership (Figure 2.3).

Figure 2.3: Land fragmentation is a key factor influencing agricultural sustainability (*Source:* https://www.ucsusa.org/our-work/food-agriculture/solutions/advance-sustainable-agriculture#.Wow2rryWbcs).

FAO (2003) recognizes the challenges enforced to agriculture and the food production by the diminishing sizes of farmland. During an International Federation of Surveyors Conference at Prague in 2000, the processes could arise as a result of cultural and legal traditions in form of inheritance and gifts or due to land scarcity. Violeta (2000), in her presentation at the 94th EAAE Seminar observed, "Land fragmentation is an effect of transformation from one system into another and has adverse effects on agricultural production." McNeill, Nesheim, & Brouwer (2012) affirmed that land fragmentation is a sign problem, which seriously affects rural development in terms of productivity, efficiency and competitiveness for farmers.

Fragmentation of land results from the combination of different factors. In most developing economies, population growth is the major cause of land fragmentation, as the availability of arable lands decreases. This affects agricultural production, making it a nightmare for many households to produce adequate food to feed their families as they have tiny plots. Most households are known to eat all the food from the farm before it is ready for harvest, thereby subjecting themselves to a long wait until the next season. This is due to the fact that they heavily rely on agricultural lands to feed themselves, and in the absence of enough food, they tend to suffer significantly.

McNeill, Nesheim, & Brouwer (2012) noted "land fragmentation has emerged as a side effect with detrimental implications for sustainable economic growth and social development." This was affirmed by Raghbendra (2005), who reported that generational succession in India had played a key role in perpetuating the cast system, which in the process reduced the size of land holdings. This greatly affected agricultural productivity. Osei (2009) affirms that Ghana is expected to increase by 45 million people and its average farm size will reduce from 0.25 hectares to 0.09 hectares per capita. This translates to more fragmentation of land. Agricultural productivity in Nepal has been quite low and decreasing. Both production and productivity were found stagnating. Research has shown that there is an increasing trend in the number of land parcels and a decreasing trend in the size of land parcels primarily due to hereditary tradition of equal division of land among the inheritors. Africa has considerable arable land but highly fragmented due to the growing population size and the culturally determined inheritance practices.

In Kenya, the pressure for farming space aggravates the high rate of land fragmentation. Land fragmentation in the country's arable land areas is closely tied to cultural values and practices associated with land ownership as well as territorialism practiced by ethnic groups (Omosa, 1998). The other stems from the colonial period when some land was alienated for large-scale expatriates and government land. The practice of land subdivision between generations has contributed to extreme land fragmentation and increasing landlessness in a country where there is no clearly defined or codified National Land Policy on the minimum acreage. Pressure on land in areas of high density has made the average land parcels economically and socially unviable. Omosa (1998) argues, "The subdivision of farms means that many of them are too small to provide the minimum subsistence to feed the large population." The tightly spaced land plots constrain sustainable food production at local and national levels where a smaller area of cropland results in less food per person on average.

2.6. TECHNOLOGICAL FACTORS

The answer to a country's agricultural productivity is not the importation of food but sustainable food production. A better understanding of the driving factors and how they impact on food production is urgently needed. Countries such as Ghana, in Africa have succeeded in Genetically Modified Organic crops (GMO) crop production measure against hunger whereas mixed cropping systems and agroforestry are some of the ways used by various countries to improve agricultural sustainability and productivity.

Many countries, especially in developing countries are behind in adapting to new technologies of agricultural production especially regarding the growth of GMO crops, where fear establishes as a side effect, though yet to be manifested. Economists such as Borlaug suggest that each country should focus on the production of commodities in which it has a comparative advantage. In the wake of change of attitude to embrace biotechnology in the rural areas, sustainable food production might just be a reality, whereby the farmers will grow for both domestic and commercial purposes (Figure 2.4).

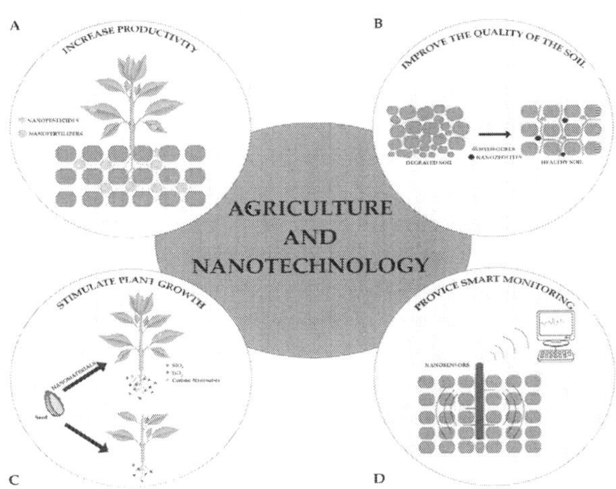

Figure 2.4: Illustrating an example of technology application in agricultural sector "(A) Increase the productivity using nanopesticides and nanofertilizers; (B) Improve the quality of the soil using nanozeolites and hydrogels; (C) Stimulate plant growth using nanomaterials (SiO_2, TiO_2, and carbon nanotubes); (D) Provide smart monitoring using nanosensors by wireless communication devices" (*Source:* https://www.frontiersin.org/articles/10.3389/fenvs.2016.00020/full).

According to National Research Council (2010), high population levels will lead to resource-depleting effects, at the expense of future generations, and agricultural sustainability will obviously be an issue. The Millennium Development Goals (MDGs) were global strategies whose aims among other things is to halve hunger and poverty by 2015. The goals were supposed to be realized through increasing agricultural productivity of crops and livestock in most countries.

Imploring the sharing, spreading and even transformation of agricultural technology are vital elements that influence sustainability in agricultural productivity. In the recent past, emphasis has begun to shift to technology dissemination more than before. The process of knowledge dissemination involves interconnected elements, drawing on existing knowledge repertoires and absorbing new information. The newly introduced information and its sources are judged acceptable and useful or contested. In some instances, farmers may already have ideas about what they are told. In this case, they may not contest the information but it may well act as a reminder or explanation of their current practice (Jose, 1989).

Many farming problems cannot be solved by science alone. Local people should be involved at formal and informal levels. Knowledge exchange and transfer is not a mechanical process. It is a social phenomenon worth investigating.

Information transfer involves interaction and is context-determined. In addition, interpretation, translation and representation of information are social acts, therefore, this calls for participatory learning process. Technologies or scientific agricultural methods should be incorporated into the local setting by involving local farmers (Christen, Squires, & Hudson, 2010).

Because these scientific methods alone cannot provide answers to all local farmers priorities and constraints, a broad approach must be adopted, one that will use and include local resources in terms of knowledge and capacity.

Several factors affect the adoption and use of technology in sustainable agricultural practice; according to UNEP (2003) these include technology transfer, capacity building, capital investment, budgeting and procurement, standards, ecosystem integrity, risk management, political considerations and stakeholders' involvement.

The spread of agricultural knowledge in rural areas is different from urban areas where communication through the media is more efficient.

When professionals visit rural areas, they are likely to meet people who are accessible and visible. For instance, these would include "people on regular salaries rather than people depending on agriculture; farmers rather than laborers; people with access to off-farm employment rather than those dependent solely on cultivation; men more than women or women more than men; those in project areas rather than those outside; users of services rather than nonusers; those who go to meetings rather than those who stay at home; those who go to market (who have something to sell, or something with which to buy) rather than those who do not go because they have nothing; those who are alive not those who are dead;" (vanLoon & Hugar, 2005).

Therefore strategies to improve local agricultural production should first involve a redefinition and improvement of roles and responsibilities for all types of institutions and individual farmers. In addition, strengthening research-extension linkages and fostering a common participatory approach is vital and, finally, the capacity and empowerment of local farmers should be strengthened (Ragasa, Catherine, & Mariam, 2016).

It is not easy to standardize ways of interacting with rural farmers. Likewise, it may not be possible to standardize dissemination techniques that apply everywhere equally. In recent years, there has been a blossoming of participatory approaches to research, extension and planning institutions by government and NGOs. This great diversity, especially in dissemination of agricultural technologies is a clear sign of strength and development. Implying that each variation is to some extent dependent on contexts and challenges specific to locations and institutions (McNeill, Nesheim, & Brouwer, 2012). The approaches are flexible enough to be adapted to suit each new set of conditions and actors, and so there are multiple variants.

According to Wright (2012), there are several barriers that hinder the use of appropriate technology so as to foster agricultural sustainability. Some of these barriers include;

- Social-political barriers such as lack of awareness and social-cultural influence.
- Technical barriers such as lack of local expertise and locally available farm appraires.
- Economic barriers, such as affordability and reusability.
- Environmental barriers such as resource consumption and conservation as well as the need to comply with environmental standards.

To address the barriers identified above several steps have been suggested as important in making sure that technology being used in agricultural activities is sustainable. Jose (1989) proposed six crucial steps that are important in ensuring that technology contributes to sustainable agricultural practices.

First is relationship building in which the focus should be on establishing cordial relationship with the local farmers before introducing new technologies so as to foster acceptance and also in building local capacity. This is done by involving the local farmers at all stages including in the identification and meeting of needs; in planning; and in technology selection, design and construction; as well as in evaluation and in the provision of ongoing technical support.

Secondly is the need for planning in which assessment of a farmer's technical, socioeconomic, and institutional needs, aspirations and resources are done as an essential component in technology transfer. Similarly, it is in the course of planning phase that potential barriers to adoption can be identified.

The third step is creating awareness, since awareness is a necessary precondition if a technology is to be promoted, funded, implemented and adopted. Awareness thus helps in reminding potential recipients of the importance of the technology being adopted as well as its effects on the environment.

The fourth step is selection; this involves the choice of technology, its adaptation to local conditions, detelimitation of the scale of the project, and identification of the recipients. Omosa (1998) assert that selection should take into consideration the ethical and knowledge degree upon the community.

The fifth step is implementation in which the establishment or construction of the technology is done while ensuring that beneficiaries are educated on the use and maintenance.

The last step is evaluation which is a process that should begin right from the initial stages during evaluation where primary operational needs and means of ensuring better tailoring the adoption process are identified.

2.7. INFORMATION ACCESS

Studies have shown that farmers who have access to technical knowledge normally experience higher production than those without. It was also noted that farmers in partnerships receive services such as vaccines, vitamins, extension and technical assistance from farm partners.

Available global trends indicate anticipated potential of increased market with estimate that between the years of 1997–2020, the consumer demand for milk, eggs and meat will increase by 50%.

Factors which will lead to the increased demand are: increase in income levels to 2USD per day which will lead to increase in demand for livestock products; and livestock production growth shifting to peri-urban areas due to ease of access to the market.

Current trends show that poultry farmers tend to be located in Urban and Peri-urban areas in developing countries due to ease of market access. Pearson (2011) says that with efficient marketing information system, poultry can be practiced in all areas due to predictability and ease of communication through use of appropriate ICTs. The information access infrastructures need to be developed and used by all players to take advantage of the great potential in increasing food needs. Only efficient systems will benefit from the anticipated global opportunities.

According to Veenhuizen & Danso (2007), ICT has contributed to increased education level and by implication income levels around the world due to enhanced communication and information access. In other words, ICT has facilitated communication and vital knowledge transfer across continents and countries benefiting all sectors including agriculture, whereby farmers have used the internet to get their products in international market. Diverse farming systems such as vertical farming technology being practiced in urban and peri-urban livestock fanning increases chances of commercialization of such species as poultry. This is a common practice being implemented in urban farming setup at various levels of sophistication (Veenhuizen & Danso, 2007).

2.8. SOCIAL-CULTURAL FACTORS

Agriculture cannot occur in isolation; it should consider the social setup of the community to enable it to meet its objectives and thus be sustainable. Social factors also known as 'social capital' are an integral component in sustainable agricultural development. Social capital provides a means through which one can measure how well a community can cooperate to achieve a desired result. According to Bhullar (2012), the lack of trust and non-reciprocity cooperation is uncommon in many communities. He argues that without this requisite level of trust and cooperation, indigenous efforts in community development are constrained and cannot be sustained. It is possible to conclude that the success of agricultural sustainability is dependent on the level of mutual trust among the existing networks in the community.

Several factors have been identified to influence the social dimension of agricultural sustainability, hence diminishing the social capital. Some of these factors include poor government policies and development policies that increase the social imbalance by worsening ethnic, class, gender, cultural, and gender tensions (Bhullar, 2012). He, therefore, argues that for agricultural activities to be sustainable, new networks and norms must evolve endogenously.

Communities are based on social relations between people who begin along lines of families and clans sharing similar values and practicing same vocation. According to Trigo (1991), this kind of linkage is important in the sense that the way communities organize themselves to tackle problems determines how well they are likely to participate in agricultural projects, even though it has been realized that the extent of kindship in the families does not necessarily reflect active participation in community development.

Johnson & Scholes (2002) argue that community networks make a big difference when it comes to enhancing sustainable agricultural practices. They assert that communities that socialize themselves around its traditional networks and activities are more likely to establish agricultural practices as their highest priority. Change in cultural facets or traditional norms also affects community participation in agricultural development activities, for example, Rao & Ibanez (2005) argue that when traditional cultural practices, fashions and vestiges that hold the communities together change, community cooperation and interest in agricultural development is also likely to decline since the burst in enthusiasm for collective action displayed by the community declines. Communities rely on their cultural identity to define themselves hence agricultural sustainability and development should take into considerations the cultural preferences of the community as they attach both sentimental and esthetic value to them.

2.9. REGION FACTORS

Religious beliefs and practices are one of the ways in which communities define and organize themselves socially, the beliefs that people subscribe to will affect their participation in agricultural activities in one way or another.

Rao & Ibanez (2005) in their analysis of the social impact of social funds in Jamaica realized how important religious organization are in promoting agricultural sustainability; they concluded that higher levels of community participation in development projects was high among groups that identified themselves to be people of the same faith, for example, Protestants, thus emphasizing the important role of the church in Jamaica's communities. Dorsener (2004) when exploring social exclusion and participation in community in Senegal realized how crucial religion is towards agricultural development at the community. She states that when one religion is dominant in an area the less dominant religion is less likely to be involved in projects initiated by the dominant community as is the case of Senegal where 95% of the population is Muslims and only 5% are Christians. She further explains that this affects acceptance of projects by communities since already there is exclusion of certain people in early decision-making phases. She proposes that adequate consideration and discussion with religious leaders is necessary if collective participation in community projects is to be achieved.

Religion can have a positive effect in ensuring sustainable agriculture in many ways. First, religious groups subscribe to some form of ethics whose focus is to ensure that needs of the impoverished are taken care of. Furthermore, these ethics are important in maintaining social values that can help in promoting good health in the community, for example, churches advocate for abstinence from pre-marital sex which can help in reducing the spread of HIV/AIDS, thus maintaining a healthy work/labor force to provide the manpower needed in agricultural productivity. Secondly, religious groups or churches provide means through which people in the community can associate with one another thus it can be used to enhance people's participation in agricultural projects ensuring agricultural sustainability (Tikjoeb, 2004).

She also asserts that churches or religious organizations provide shelter, food healthcare and education to the marginalized which can be used by the government and communities as a means to promote a working partnership that can help in ensuring success in implementation of sustainable agricultural projects.

2.10. CROPPING PATTERN

The way in which agricultural crops are planted in terms of spacing and mixed cropping has a certain effect on agricultural crop yield. Trigo (1991) points out that productivity is a function of crop mix. According to Trigo, intercropping varies from place to place and, apart from influencing productivity, it also results in equally marked difference in annual labor inputs.

The fact that the cropping pattern has a significant impact on yield and labor engagements means that it also has implications for rural incomes and its variation from place to place.

Sound cropping pattern for each crop, therefore, becomes a useful consideration in any attempt to bring agricultural sustainability that will have a meaningful impact upon incomes, food security and agricultural employment.

The paradox is however that while such unsound cropping pattern has the said adverse bearing on farm production and therefore incomes, the small-scale farmers have often resorted to the same farm practice with an aim of maximizing, within a given season and farm holding, their crop yields and subsequent incomes.

2.11. STORAGE FACILITIES

Storage facilities for agricultural products are very important as they highly influence agricultural sustainability in one way or another. The ability of the agriculturalists to store safely the excess produce until it´s required or marketed, avoiding it gets destroyed by weevils and other pests, as well as environmental conditions, are crucial. This ability will have a decisive bearing on their willingness to produce more crops on top of their immediate subsistence requirements. It will also have an influence on marketing and pricing as farmers without proper storage capacities will hurry to the market and flood it with crops which they fear might get spoilt due to their inability to properly store (Bluffstone & Gunnar, 2012).

Since most of the farm produces are perishable and prices tend to be at their lowest during harvest times, the farmer who has no access to proper storage facilities suffers significant loss of income. Therefore, it is always important to assess the exact extent of the storage problem and establish the magnitude to which crops get wasted due to this and also the extent to which the absence of such facilities or such wastage offer an explanation to the low level of agricultural production. This is because many governments have often missed the gun by tying unduly large amounts of capital to huge stores which have become white elephants because they operate at negligible capacities due to their over-estimation of the magnitude of the storage problem and the misguided contention that this will be an incentive to farmers to step up their agricultural production (Acton, 2013).

It is important to find the precise nature and magnitude of the losses, differencing between the different stages of marketing, crop, areas and costs. This information will provide greater insights into determining the real cost of storage construction confronting real savings and real losses.

The root of the matter is that farmers cannot save on the losses of the surpluses which have not been produced; the first concern should, therefore, be to design avenues through which such excess produce can be forthcoming to warrant the concern about storage facilities.

2.12. AGRICULTURAL COOPERATIVES

Agricultural cooperatives have several responsibilities in agricultural sector. The following are some of their responsibilities:

- Collecting, processing, transporting and marketing the product of the member farmers.
- Supplying agricultural inputs, plowing and provision of seasonal credit.
- They diversify into commercial savings and credit functions.

According to Macnaghten & Ripalda (2015), "Developing agriculture for increased agricultural production needs production inputs. The channels for supplying these inputs are not well established mainly because the traditional expenses of administering a multiplicity of small loans to small-scale farmers and low productivity and the high risks involved tend to discourage the ordinary bank from entering the field. The increased production financed with these loans enables the farmers to save through their societies and this finances their longer-term farm investments."

Zhen & Zoebisch (2006) note that "Today with the very pressing problems of worldwide food shortages, co-operation has become part of the 'package deal' by which governments are trying to lift their agriculture out of its subsistence husbandry and techniques into the required pattern of expanding output, rising incomes and more business-like attitude cooperation is part of the governments' administrative machinery for securing the fulfillment of national plans, and it must be judged in this context. Agricultural cooperatives must therefore command a central place into our agricultural systems in their drives to expand and develop the agricultural sector."

2.13. POLICY FORMULATION

Studies show that agricultural performance and sustainability are affected by so many problems. Therefore, policies related to these factors must be rationally designed and incisively articulated. Practicability and efficacy should be the prime and constant factors behind such formulation.

If a lack of proper agricultural policies is encountered, agricultural sustainability is likely to be severely affected. This is because farming is riddled with so many problems that if there is no deliberate effort to encourage it, investors would opt out of it. Therefore, agriculture must remain in the forefront in the national policy making decision-making process (Acton, 2013).

There are several reasons that can explain why most agricultural policies fail in one way or another. Some of these reasons include:

- Many agricultural policies development are often formulated against the background of inadequate, unreliable and at times irrelevant field data. Therefore, most of these policies fail to really address themselves to actual and existing problems.
- Agricultural policies have often failed because their articulation has never taken cognizance of the world-views and situational realities of the people.
- Policy decisions often ignore the small farmer's views. An agricultural policy must be judged in the context of the cultural and social environment of which it is part.

There is often a big variance between stated policy prescriptions and what is actually affected.

2.14. MARKETING AND PRICING

Agricultural sustainability is influenced by the pricing and marketing of agricultural products.

Marketing and pricing help in introducing and sustaining change in the agricultural sector, and this ensures increased production. This is because marketing and pricing provide major incentive systems that will influence the enthusiasm with which farmers will take on agricultural practice or accept the introduction of new crops or even new technology that entails increased production and therefore disposable surpluses through the market.

But often price controls related to management of floor and ceiling prices for crucial crops as well as market access are not effectively handled. Providing the grounds to farmers to grapple with stifling price fluctuations and the uncontrolled local markets which are more useful to such farmers but whose operations are closely related to those of black markets (Acton, 2013).

It is a common reality in rural areas that prices in such markets often hit rock-bottom and rocket sky-high at very wrong times for the small farmer due to the absence of effective price policies and controls relating to the very local markets for the very small farmer.

According to Nair (2016) argument, "Government marketing systems have been set up in many countries, but their benefit to the producer has been questioned. Even in cases where a government makes a genuine attempt to procure a fair deal for the producer, there is a tendency to set up systems in which overheads are so costly that the real value of the crop cannot support them. The result is that the margin between payments to farmers and the market price can be as high as 50%." Pricing policies according to Nair (2016) have also tended to act against the farm output. "Governments influenced by the power of the urban/industrial sector see cheap farm produce as one way of transferring resources from agricultural sector to other sectors in the economy."

There are several factors that affect the marketing and pricing controls in agricultural sector. Some of these factors include: inadequate competent personnel; inadequate and imprecise relevant data related to marketing and pricing together with institutional weaknesses; and poor knowledge of the size and sources of food production and marketable surpluses. These issues need to be handled both at local and national level in order to enhance agricultural sustainability in a country.

2.15. AGRICULTURAL EXTENSION SERVICES

According to Agricultural Extension Society of Nigeria (2015), "agricultural extension service is an evolution of the advisory service which can be regarded as a form of community development with an agricultural bias and an educational approach to the problems of rural communities." The willingness and ability of skill acquisition among farmers usually has great influence on agricultural productivity and sustainability. This depends on both training and economic incentives. Such services become even more critical for agricultural improvement in a rural setting where the level of literacy among small-scale farmers is still considerably low. It is however important to first understand the underpinning principles of effective extension work.

Omosa (1998) says that in the early colonial departments of agriculture, advisory services were established to explain to farmers what they should do to increase output from their holdings. In order for them to realize the government's objective of increased agricultural production. These advisory services, he rightly says, were often wrongly referred to as 'extension services.' They were not extension services really but more directed towards the educational sense of the term. They were designed to advice farmers on improved farming techniques which would help and assist them to implement a benevolent government's plans for the development of the country's economy (Omosa, 1998).

The success of any extension work lies in a recognition and prioritization of the farmer's point of view as well as developing his confidence in himself and ability to initiate and manage agricultural change using mostly his material and manpower resources with a declining injection of government resources into such change.

Dillard, Dujon, & King (2008) stated that the primary objective is to change people's outlook towards their difficulties and develop themselves to permanently sustain change. They are taught to increase their knowledge, to make the best use of their faculties and skills and develop their capacity for work. Most of the implemented efforts have failed because the farmers cannot relate the effort to their own view on what is needed, their aspirations, sense of economic viability, and their own farm resources. Ragasa, Catherine, & Mariam (2016) say what is important is "the examination and documentation of actual farm problems over a range of farm systems and, within a given system, over different types, sizes and capabilities of farms. These must in turn become the facts of life with which extension agents and technological and economic research officers must work." They further observes that the extension officer should understand the present technology and farming systems as well as the problems of the small farmer because given the pressing need to raise per-farm productivity, it is essential that farm systems be devised which are economically attractive to the farmer and which are not incompatible with his own resources (particularly labor and cash).

There are five essential procedural steps that every extension officer should perform. These include:

- Conducting a study or survey of the conditions in the community.
- Preparing the extension program.
- Preparing the plan of work.
- Preparing the calendar of work.
- Conducting an evaluation and measurement of progress.

The inability of the agricultural extension officers to effectively steer change in small-scale agriculture could be due inadequate training which makes them unprepared to grapple with some of the very technical field tasks which otherwise should fall within their control. Their competence should be improved in handling all aspects of the farm technology package (Ragasa, Catherine, & Mariam, 2016).

3 CHAPTER

IRRIGATION PRACTICES AND AGRICULTURAL SUSTAINABILITY

CONTENTS

3.1. Introduction..48
3.2. Place of Irrigation in Agricultural Sustainability..48
3.3. History of Irrigation 52
3.4. Methods of Irrigation 54
3.5. Some Factors that Affect Crops Response to Irrigation...58
3.6. Problems Facing Irrigation Schemes In Agricultural Sustainability 59
3.7. Key Concepts of Irrigation for Agricultural Sustainability..61
3.8. Socio-Economic Aspects of Irrigation...75

3.1. INTRODUCTION

This chapter is mainly related with evaluating the impact of irrigation schemes influencing the sustainability of agriculture. The main objective of the chapter is to add to the existing knowledge on irrigation schemes a more specific technique of identifying the factors that facilitate or inhibit the efficiency of irrigation schemes in a country. There are several areas concerning irrigation practices that have been discussed in this chapter. Some of these areas include: place of irrigation in agricultural sustainability; history of irrigation; methods of irrigation; some factors that affect crops response to irrigation; problems facing irrigation schemes in agricultural sustainability; key concepts of irrigation for agricultural sustainability; and socio-economic aspects of irrigation.

3.2. PLACE OF IRRIGATION IN AGRICULTURAL SUSTAINABILITY

The importance of irrigation in the world today cannot be overemphasized. The need for survival and the need for additional food supplies require a rapid expansion of irrigation throughout the world. Although irrigated agriculture has developed most extensively in arid regions where natural precipitation is inadequate for the production of many crops, it is becoming increasingly important in humid regions as well. Irrigation has been defined by Kang (2010) as the application of water to the soil for the purpose of adding water to the soil to supply the moisture essential for plant growth, providing crop insurance against short-duration drought, tooling the soil and atmosphere, thereby bringing about a favorable environment for plant growth, and washing-out or diluting the salts in the soil and softening the tillage pan.

This definition does not restrict irrigation practices to the arid and semi-arid regions. In fact, some of the most profitable irrigated agriculture in the world are located in areas normally thought to have sufficient rainfall (Cantor 1967). These are areas such as Central Brazil, Central America, the West Indies, and the western part of Africa as well as parts of South Africa, which have extensive annual rainfall during 6 months and then face another 6 months with practically no rainfall.

Other areas have short periods of drought which require irrigation, if a profitable and diversified agriculture has to be practiced. Irrigation in such places is only supplemental to natural precipitation. Therefore irrigation is no longer a regional practice of arid and teal-arid zones but has become a basic necessity of a well-developed agriculture.

In many countries, the development of agriculture is dependent on irrigation. In others, irrigation is a prerequisite to higher productivity of limited land resources, production of high-value cash crops, and diversification and intensification of agriculture. Irrigation development can have a considerable impact on the overall economic development in terms of increased employment, increased rural incomes, and hence increased living standards which in turn stimulate the development of the internal market for industrial products. Maximum agricultural production cannot be achieved without adopting modern techniques and services such as irrigation. Even on land which has been farmed for centuries the introduction or improvement of irrigation practices can provide the exciting promise of higher yields and better living conditions for families (Esteve, Brebbia, & Ricc, 2008).

Irrigation has converted deserts into productive agricultural lands thus attracting large populations and new centers of life and civilization. For example, irrigation from the Nile is a source of food, life and prosperity in Egypt, those entire cultivated areas are dependent on the water supply provided by irrigation (OECD, 2006) (Figure 3.1).

Figure 3.1. Illustrating irrigation canal systems along the Nile River in Egypt (*Source:* http://iteachteched.net/Riemer/Shaikh212.html).

Three-quarters of the cultivated area in Japan is irrigated and the main crop is rice. Other countries which have a high proportion (about two-thirds) of irrigated farming area are Afghanistan, Guyana, and Taiwan. In these areas where agriculture is almost entirely dependent on irrigation, canals, dams, and wells are viewed not only as the backbone of the people's economy but also as intimately related to the very life of the people. In this regard irrigation is very much a human problem. Lack of it would mean devastation and misery for large parts of the population (Pereira, Feddes, & Gilley, 2013).

Rice, which is the staple food for more than half of the world's population, is grown, except in small zones of very heavy rainfall, entirely as an irrigated crop. Similarly, good quality cotton cannot be successfully grown without adequate irrigation. Fruit and vegetable production are eminently enhanced by irrigation. When there is insufficient food for people in a country and, for various reasons, it is difficult to import it; the magnitude of costs of an irrigation project may be regarded by the authorities as relatively unimportant.

Irrigation development not only helps in increasing food production but also opens up new land frontiers to alleviate the population pressure in densely populated areas, which is one of the development problems in many developing countries. The need to create more reliable water supplies to be able to use a large proportion of seasonal river flows and reduce floods has given impetus to the construction of storage dams. Another factor contributing to this need has been the opportunities for developing additional water supplies through simple diversion that have been exhausted (Kang, 2010). However, the cost of these projects, their long gestation periods, and the relatively high foreign exchange component in their construction have caused some economic problems when using them for water-supply development in many developing countries. Their optimal use requires a high level of farm management not only for the water supplies created, but also in their utilization at farm level. The same need for optimal utilization equally applies to all water development for irrigation. An optimal water development plan should be oriented not only towards developing new-water supplies, but also to increasing the efficiency of use for the available water.

Strategies for the use of latent water resources will depend to a large extent on the costs of development and the potential productivity. While surface water resources have provided the base for the bulk of irrigation in most of the world's irrigated regions, groundwater has played and will continue to play an important part in irrigation development especially in areas like India, Pakistan, and parts of the near East and Africa (OECD, 2006) (Figure 3.2).

Exploration of underground waterresources for irrigation purposes

Utilization of underground water resources in irrigation schemes

Figure 3.2: Illustration exploration of groundwater and its utilization in irrigation schemes for sustainable agricultural practices (*Source:* https://www.atlantismuhendislik.com/en/urunlerimiz.php?pid=22).

Recent trends have shown a growing awareness of the utility of groundwater and technological development regarding drilling and pumping which have brought opportunities for their wider exploration. Most of the world's smaller and/or easier irrigation schemes based on river capture, especially in the developing countries have been avoided because of the large capital investment requirements. Research in India has revealed that exploitation of groundwater is generally less expensive than that of surface water. However, location of some groundwater resources are not known with any marked precision (Tejero & Zuazo, 2018).

Development costs have been estimated by analyzing a wide range of agricultural development projects assessed and/or financed by the international bank. From this, it has been possible to determine the likely range of development costs for different types and sizes of irrigation projects.

A general examination of these costs suggests that no economies of scale are likely to occur. The irrigation development costs per hectare have been found to increase linearly with the size of the project (Pradeepkumar, 2008).

3.3. HISTORY OF IRRIGATION

According to FAO (1970), approximately 13% of the cultivable land used for annual and permanent crops in the world was under irrigation in 1963. This was approximately 160 million hectares. By 1975, this area was estimated to be about 200 million hectares and before the turn of the century the total hectares had exceeded 300 million hectares. The developing countries alone commanded over 162 million hectares of irrigated land in 1975 although almost half of it requires rehabilitation or improvement, and much of the available water-flaw is underused (Table 3.1).

Table 3.1: Leading Countries in Irrigated Land

No	Countries	Area irrigated in million hectares
1	Australia	1,150
2	Botswana	2
3	Brazil	141
4	Burma	753
5	Canada	627
6	Ethiopia	30
7	France	2,600
8	India	37,640
9	Indonesia	3,797
10	Iran	4,000

Source: https://www.slideshare.net/raiuniversity/b-sc-agri-ii-wmmi-u-1-irriga-tion.

India has the largest irrigation schemes in the world outside Mainland China. Among the leading regions in irrigation are Asia and the Far East, North America, U.S.S.R., North West Africa, Europe, and South America. Irrigation is particularly important in the Near East where the agricultural systems are based on the Nile and the Tigria-Euphrates, and where agriculture would be impossible without irrigation.

The water resources of a country are its rivers, lakes, and springs refilled by rainfall and under-the-surface sources which usually developed millions of years ago and essentially are available for "mining." Accordingly, different types of irrigation works have been developed over the past ages, examples of which are:

- Percolation wells (Artesian wells and tube-wells).
- Tanks (earthen storage areas).
- Large storage reservoirs.
- Pumping or lifting from rivers and lakes.
- Weir-controlled diversion canals.
- Trans-basin diversions.
- Multipurpose projects.
- Different combinations of all of them.

Percolation wells as a source of irrigation waters goes back to prehistoric times and is still popular in many parts of the world. In India, alone, about 5 million wells are in use for irrigation. A large number of tube-wells have been installed in many areas of the world in the last 30–40 years.

Tanks and storage reservoirs have been an important source of irrigation supplies for ages, mainly in Ceylon, Continental China, and Central and Southern India. Lifting water from rivers is also an early means of irrigation for areas along the riverbanks. Where irrigable areas are low, lift may be necessary only in the season of low flows. During the high flow period, in such cases, river water can flow to irrigate areas using and benefitting from gravity. The source of power for lifting, in general, may be manual, animal, thermal, or electric. Thermal power includes natural gas and petroleum.

In hilly areas, where river channels are relatively steep, it is common practice to divert streams of water into small channels taken along the hillside for purposes of irrigation (Figure 3.3).

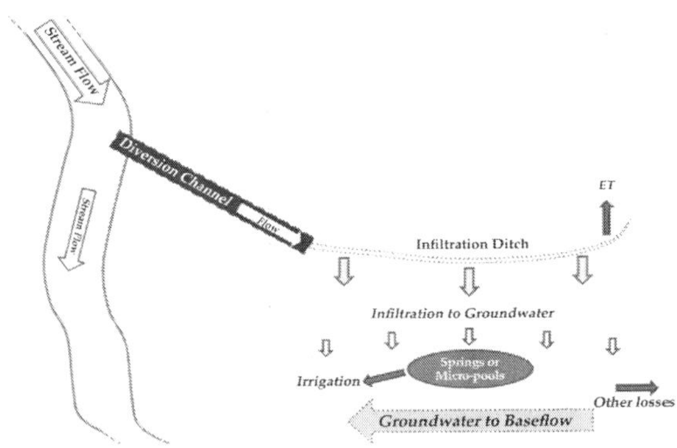

Figure 3.3: Diversion Channel from mountain streams to shallow small channels for irrigation purposes (*Source:* http://www.watershedconnect.com/news/

to_avert_disaster_peru_diverts_brwater_fees_to_andean_alpacas_and_prein-can_aqueducts).

During the last 150 years, a large number of weir-controlled canals have been built in different parts of the world. These canals are based on the run of the river, with no storage. These may be l-seasonal, supplying water for one crop season only, 2-seasonal, or perennial.

With the advance in the knowledge of hydraulics and the development in engineering techniques during the last hundred years, many large-size storage reservoirs have been built by damming the flow of rivers by masonry dams (30 to 100 meters high), concrete dams (up to 284 meters high), and earthen dams (up to 240 meters high). In conjunction with canals, these reservoirs provide Irrigation to large areas (FAO/UNESCO, 1973).

3.4. METHODS OF IRRIGATION

For economical and efficient distribution of irrigation water, the operator must at all times have complete control of the water as it flows from the head ditch onto the land. This is true, whatever method of irrigation is used. When uncontrolled streams of water are turned into the fields, waste, inefficiency, and uneven distribution are almost certain to result. This can be averted by the use of relatively simple equipment enforcing a means of distribution and control of water (Kang, 2010). Irrigation water is applied to land by three general methods namely:

- Surface application by flooding.
- Sub-surface or with furrows in which the surface is wetted little if any.
- Sprinkling, in which the soil surface is wetted as much as it is by rainfall.

The surface and sub-surface methods are further subdivided as follows:

1. Surface application

- Uncontrolled or wild flooding.
- Flooding controlled with corrugations, borders, and basins.
- Furrows.

2. Sub-surface application

- Controlled by lateral supply ditches.
- Uncontrolled irrigation through excess application of water to adjacent or higher lands

Irrigation methods vary in different parts of the country and even on different farms within a community because of differences in soil topography, water supply, the crops grown, and the custom. Close-growing crops are normally irrigated by use of borders and basins.

Some vegetable crops are all suited to flood irrigation by borders and basins. Row crops generally are irrigated by furrows. Any one or a combination of several methods may be best suited to one farm. Although sprinkler irrigation is used by a few farmers whose farms are very uneven, the most widespread method of irrigation in most parts of the world is by flooding, especially basin flooding and furrow irrigation.

Sandy soils usually have too high an intake of water. In view of this, they are best irrigated by over-head means, but this method is far more expensive than flood irrigation. The essential requirements for flood irrigation are sufficiently smooth land of very gentle gradient; preferably flat (Tejero & Zuazo, 2018). Where flood irrigation is being carried out there must be complete control of the water and the farmer must know the amount of water his layout is capable of applying. Uncontrolled irrigation leads to overwatering, inefficient water-use, and hence poor crop returns and inefficient water management.

Flood irrigation is generally the simplest and cheapest method of applying water. However, it has its own disadvantages requiring constant attention towards maintaining levels and smoothness. More skilled labor is required to apply water evenly and to avoid waste through excessive runoff than for most other methods. The canals require constant cleaning and maintenance (Figure 3.4).

Figure 3.4: An example of flood irrigation scheme (*Source:* https://commons.wikimedia.org/wiki/File:LevelBasinFloodIrrigation.JPG).

The one great advantage of a flood scheme is that it is very flexible. If necessary it can be adjusted easily by enlarging the volume of water to increase the acreage irrigated or to irrigate the lands more quickly than originally designed.

Row or furrow irrigation is probably the cheapest method of obtaining efficient irrigation. It is suitable only for crops which can be planted in rows sufficiently far apart to allow furrows to be made between the rows (Figure 3.5).

Figure 3.5: An example of row or furrow irrigation (*Source:* https://www.snf.us/markets-served/agriculture/furrow-irrigation-image/).

Each row may have a furrow serving it or where the rows are fairly close together, one furrow may serve two rows of plants. Irrigation water is run between the crop rows. The size of the stream is varied according to the gradient and soil texture. For flat gradients, long runs, and sandy soils, large flows of water are used (Kang, 2010). In any case, the ideal stream is of such a size as to run to the end of the furrow with inflow just equaling the infiltration in the furrow. Row irrigation is the most common method used to irrigate the banana crop, for example.

Basin flooding is widely used in irrigating fruit trees and vegetables on flat topography. Basins are flat areas surrounded by low ridges or dikes. They may be square, rectangular, or irregular in shape and may vary in size from 6 foot squared to an acre depending on the soil texture and the size of the irrigating stream (Figure 3.6).

Figure 3.6: An example of basin flooding irrigation scheme (*Source:* https://cals.arizona.edu/aes/yac/research.html).

The more porous and sandy the soil is and the smaller the irrigating stream, the smaller should be the basin for efficient irrigation. Where vegetables are irrigated, each basin is made and leveled independently of the others. The basins are filled with water to a depth of 2 to 6 inches depending on the soil type and the crop. Sandy soils require more water because the infiltration rate is higher.

Water from the source is led through the main canal, the size of which depends on the size of flow and the size of the irrigating system. From the main canal the water is diverted into an intricate canal system which distributes it throughout the farm. The most common means by which this is done is with open ditches or laterals. The ditches are generally permanent features and commonly follow boundary lines, fences, and edge of fields. They are frequently earth ditches which may suffer from excessive losses due to seepage and evaporation, especially in arid regions or in areas of porous or sandy soils.

Leading from the permanent open ditches are the field ditches which may or may not be plowed in at the end of the growing season. Water is delivered through the field ditches by means of check structures or turnouts. They usually consist of metal or wooden fixtures, though they may be merely gaps cut in the ditch bank. From the field ditches water is led into supply ditches and finally into individual basins one at a time by cutting a gap in the levels surrounding the basins. This method of irrigation does not allow heavy mechanization. It requires much human labor with a high degree of skill for adjusting the flow of water to avoid over-watering or under-watering (Ali, 2011).

3.5. SOME FACTORS THAT AFFECT CROPS RESPONSE TO IRRIGATION

3.5.1. Excessive Irrigation

Too frequent irrigation can waste, labor and limit the utilization of the crops. It can lead to losses of water and nutrients in solution through surface runoff and leaching. Dutta (2011) observed that crops yields increased with increasing precipitation up to an optimum point; beyond that point yields decreased as precipitation further increased.

3.5.2. Interdependence between Moisture and Fertility

There is considerable experimental evidence to show that tremendous wastage does occur when irrigation water is applied to crops which are severely limited in their response by inadequate soil fertility, particularly nitrogen. The reverse is equally true. According to the observation made by Snellen (1996), it was concluded that maximum response to fertilizer was not obtained if water supply was inadequate; he reported that when moisture was low and temperatures high addition of fertilizer decreased yields and that addition of either fertilizer or water alone often did not significantly increase yields but both factors. Similarly, Verma (1993) found that in dry years irrigation markedly increased the return from potassium fertilization.

Interdependence between water and higher nitrogen supply was established by Ali (2011) in their effects on the productivity of crop with lower rates of application the effects of nitrogen and irrigation water appeared to be independent. Klelnschmidt (1967) also obtained response to irrigation at 200, but lacked response between 0 to 50 lbs nitrogen per acre, which suggested that nitrogen was the first limitation to production but that water was limiting at higher nitrogen levels. The same explanation can be inferred from Ofosu (2011) who reported that in dry years, soil moisture was depleted more rapidly in the fertilized crops. It can thus be concluded that the magnitude of the response to nitrogen and the relative efficiency per unit weight applied depend upon moisture availability.

3.5.3. The Type of Plant

Research has shown that the yield of a plant depends both on its innate genetic potential and the availability of sunlight moisture and nutrients. Sunlight is abundant in the tropics, moisture is low in many areas but could be conserved, but that lack of nutrients is the most important limiting factor for getting yields from the varieties with a good genetic potential for yield (Takeshima & Edeh, 2013). This is exaggerated by the inability to make nutrients available through the soil due to the lack of adequate moisture in the soil.

Two types of crops are known, the natural plants (consisting of all kinds of plants growing on hard virgin land), and planted plants (made up mostly of selected species with desirable genetic potential and growing on tilled soft and deep soil). Dutta (2011) obtained better response to fertilizers with irrigation from planted crops than from natural crops. This was expected since water and nutrients were not only shared by weeds but also more liable to surface run-off for lack of infiltration into the hard and undisturbed soils of the natural plants.

3.6. PROBLEMS FACING IRRIGATION SCHEMES IN AGRICULTURAL SUSTAINABILITY

In view of the breakdown, in the past, of many civilizations that were based on irrigated agriculture and the numerous cases of rapid soil deterioration in modern times (both in countries with under-developed agriculture and in those with the most advanced technologies), doubts are frequently expressed upon the possibility of maintaining irrigated agriculture permanently.

Irrigation development has experienced both social and technical problems. The success of irrigated agriculture depends on the compatibility of water, land, and people (Ali, 2011). The world today has many abandoned irrigation projects, caused primarily by inadequate considerations of the combined uses of these resources. Many irrigation schemes particularly in Africa have met with many settlement and administrative problems.

Only a few irrigation settlement schemes in Africa can be termed as successful, most others having failed outright, or having operated over a long time without recovering their costs (Ofosu, 2011). Most irrigation schemes in the early stages face the problem of selection of the right settlers. Selection of tenants who are not committed to the hard work involved in irrigation may lead to instability in the early stages of implementation.

Another major problem which faces irrigation settlement schemes is the strict requirements of high-value crops in the scheme, which in many cases conflicts with the subsistence requirements of the tenants. Water sharing among settlers in a big irrigation scheme, or at times between countries can be a major administrative problem.

Experience and research have shown unequivocally that the basic causes of the failure of crop production under irrigation are the combined and related effects of excessive salt-accumulation in the root-zone and the development of a high water table (Ali, 2011) (Figure 3.7).

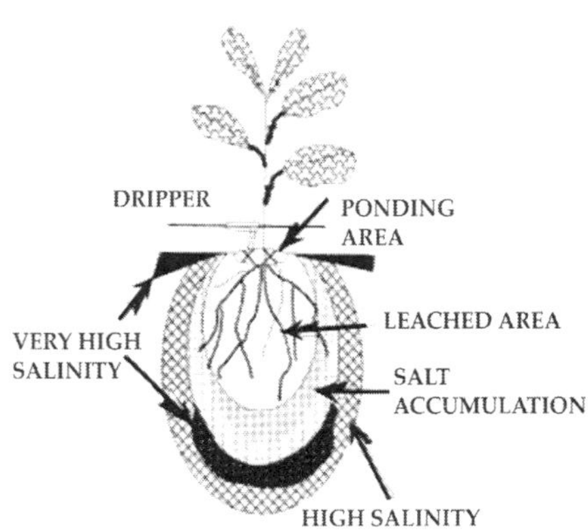

Figure 3.7: Soil, water, and salt distribution in soil with a drip irrigation system (*Source:* http://www.haifa-group.com/knowledge_center/crop_guides/strawberry/special_sensitivities_of_strawberries/).

Nevertheless, scientists agree that these problems are not insurmountable. The fact that good crop yields have been maintained under irrigation for a period of more than 400 years in both Egypt and China supports the thesis that irrigation can be a permanent feature and one of the most important and productive systems of agriculture. History and research have shown that for irrigation development to be successfully maintained, there is need for a strong central government to construct and maintain extensive irrigation schemes, proper design of the system, provision for adequate drainage to match the increased availability of surface water, and careful control of irrigation practices so that the persistent problems of erosion, waterlogging, salt accumulation, soil permeability and aeration, and soil depletion can be controlled.

3.7. KEY CONCEPTS OF IRRIGATION FOR AGRICULTURAL SUSTAINABILITY

Although in many parts of the world irrigation development has been extended to cover even the humid and sub-humid areas, in most countries, much of the irrigation development efforts have been concentrated on the so-called medium and low-potential zones.

According to World Bank (1999) suggestion, the main objective in concentrating efforts in these areas has been to mitigate the effect of drought and to bring security, stability, and prosperity to those areas producing only catch crops or no crops at all. Nevertheless, as modern cash inputs are successfully removing or reducing the effects of many limiting factors, even in areas of moderately adequate precipitation, there is an increasing need to consider supplementary irrigation to prevent moisture-availability from becoming the ceiling on yields. There is therefore a possible choice of two objectives in irrigation:

- Supplemental irrigation aimed at ensuring a more or less constant level of production even during the long dry spells between long and short rains, and
- Full irrigation designed to provide for intensification of crop production.

Among the specific advantages that may accrue from irrigation are the following:

- Maximum growth and yield traceable not only to the supply of water itself but also to the part that moisture plays in making nutrient materials available.
- Timing of maturity of crops – particularly desirable where earliness is profitable.
- Maximum table and market quality (quality of the harvested product is almost always inherently improved through irrigation).
- Utilization of land irrigation usually affords an opportunity for fuller employment of land by ensuring double or multiple-cropping.

3.7.1. Profitability of An Irrigation Scheme

Ali (2011) noted that farming with irrigation is both capital and labor intensive. Because of the high-costs involved in supplying irrigation water, only the most valuable crops can be grown profitably. In order to secure satisfactory returns from irrigation farming, the farmers must have knowledge of the important factors affecting returns. They must also be proficient managers capable of taking advantage of every opportunity to secure increased returns from their crops. For successful irrigation farming, experience in irrigation work is essential. Farmers must know when to irrigate. This is important because expensive water may be wasted through over-watering (Ali, 2011).

Secondly, it is essential that the farmers have a high degree of managerial ability and are able to accept new ideas and innovations. Response in yields due to irrigation is only realized at high levels of complementary inputs such as fertilizers and manures. Without good management practices money spent on irrigation would be money lost in the drain (Ali, 2011).

Successful irrigation also requires that a farmer has first-hand market intelligence so that he can make quick decisions on types of crops to grow, how much of each crop to grow, and in what seasons to grow them. The aim should be to grow crops that have the highest return per acre-inch of water applied.

Good irrigation farmers stagger the planting dates of various crops in such a way that the crops mature at different times. This is one of the advantages of using irrigation water. In this way, the farmers get a regular income by maintaining a constant supply of produce (Ali, 2011).

3.7.2. Plant-Water Requirements

Plants require water for growth. Water is transpired by the leaves throughout the day which is then drawn up by roots and passed out as water vapor by leaves. This evaporation helps the leaves to remain cool. Besides the transpiration of plants, the soil also loses moisture by evaporation from its surface. The sum of these two losses of moisture from the soil is called evapotranspiration (ET) (Verma, 1993) (Figure 3.8).

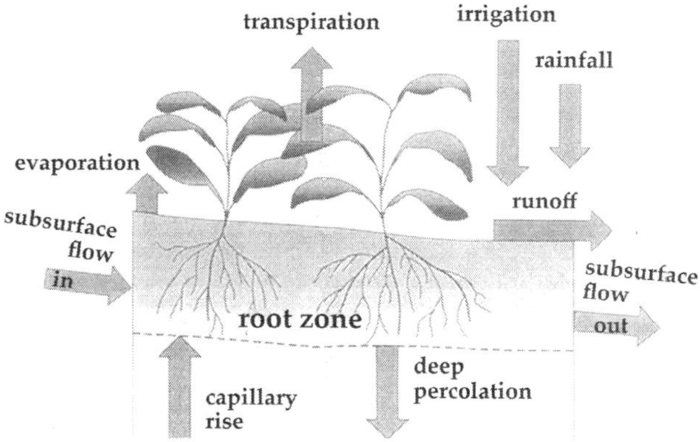

Figure 3.8: Illustrating the effect of evapotranspiration on irrigation practices (*Source:* http://www.fao.org/docrep/x0490e/x0490e04.htm).

To the surprise of irrigation engineers and farmers accustomed to thinking in terms of each crop having its own water requirement, an important conclusion of research in science by Penman (1963) has established the theorem, based on the simple laws of physics, that maximum water requirements for all crops must be about the same if the crops are grown on the same soil types, under the same conditions of temperature, sunshine, humidity and wind velocity, and for the same growing season. This maximum ET of green crops grown under the same climatic conditions is called potential evapotranspiration (PET) and is dependent only on the weather conditions and not on the nature of crops being grown (Snellen, 1996). In general, the PET of a field with plentiful moisture and fully covered by a green crop cannot exceed the evaporation from an extensive body of water exposed to the same weather conditions. This result permits estimates to be made of the daily water requirement to offset PET of all or any crop from climatological data.

Methods have been developed for using standard weather data to estimate the irrigation needs of a crop. Such methods do not, however, take into account the possibility that some plants may respond to irrigation only at particular stages of growth and may tolerate very dry soil conditions at other stages.

This important finding by Penman (1963) has led to the drastic revision of all previous ideas about the economics of irrigation. Irrigating a given area at a given time of the year, will use up the same amount of water almost irrespective of the crop which is being grown. Irrigators therefore should always be growing the crop which at that time of the year and at the prices then prevailing, yields highest economic return per unit of area and per unit of time. It is important to note, however, that not all crops transpire the same amount of water per unit area over their life cycle. Different crops may have different growing seasons during which climatological conditions and hence ET may differ. Furthermore, even crops with the same growing season and thus subject to the same weather conditions may take different lengths of time to reach full leaf cover.

Crops which establish a leaf canopy early have been found to use irrigation water more efficiently.

This can be summarized from the records and reviews on the subject by Russell (1939) as follows.

- Water is required to transport raw materials and finished products within the plant.
- It maintains the proper consistency of the living protoplasm in which the growth processes take place.
- It provides the necessary working pressure in the growing cells.
- It provides the sole means whereby plants obtain nutrients from the soil.
- It is required for photosynthesis.

Eighty percent or more of the plant's substance is composed of water, thus, if water is lost by plants faster than it can be taken up from the soil, or if there is insufficient soil moisture available, the result is loss of turgor, and the plant will droop and wilt. This forces the stomata to close reducing transpiration.

The closure of stomata openings is accompanied by a reduced intake of carbon dioxide and reduced carbohydrate production and, finally, cessation of growth, for only turgid cells can grow. Secondly, Ali (2011) has observed that high temperatures in the surface soil are destructive to the delicate and shallow roots of plants. This was shown to delay recovery when water again became available.

3.7.3. Soil-Moisture Relationship

One of the most important factors determining proper irrigation practices is the character of the soil being irrigated. In general, the texture, structure, and porosity of soil determine its water-retaining and transmitting capacity (Ofosu, 2011). In turn, these two capacities, together with the crop and depth of the root zone, largely govern the method of irrigation, the frequency of application, and the quantity of water that should be applied within each irrigation.

The water retaining capacity of soil, for the purpose of irrigation, is expressed as depth of water held in a given depth of soil. This is expressed in inches depth of water per foot depth of soil or millimeters of water per centimeter depth of soil. When soil is throughly watered, some of the water, under the influence of gravity, drains into the lower levels and is replaced by air from the surface. When drainage virtually ceases usually within two to four days, the soil is said to be at field capacity (Figure 3.9).

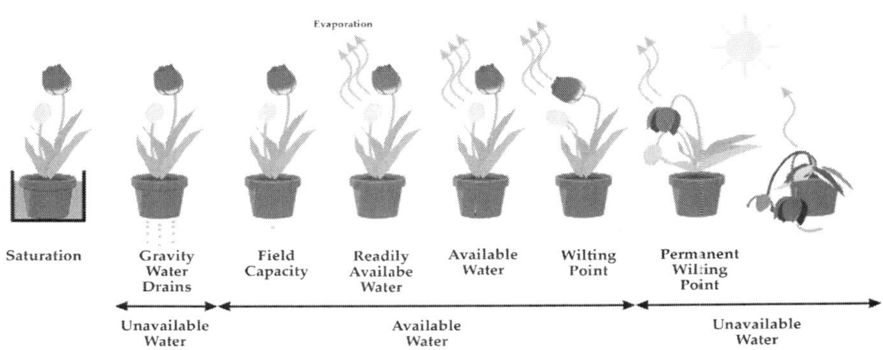

Figure 3.9: The concept of water availability for field crops (*Source:* https://soilmoisture.wordpress.com/2016/08/16/soil-water-status-saturation-field-capacity-and-wilting-point/).

The roots of crops obtain water from the film of water held around the soil particles by surface tension. As this film becomes thinner and thinner, the roots find it increasingly difficult to take in water (Figure 3.10).

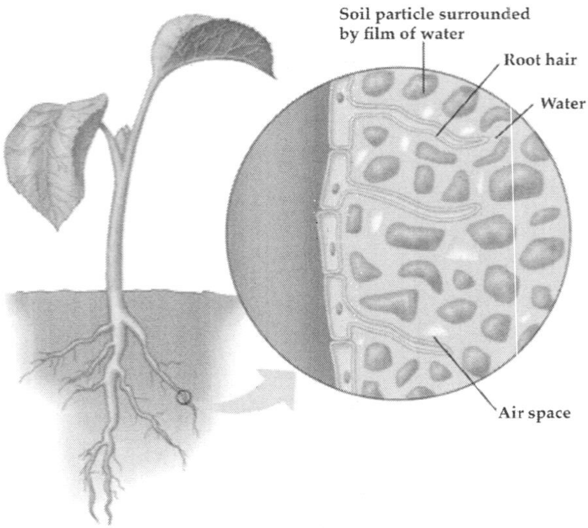

Figure 3.10: Mechanism of plant water absorption (*Source:* https://pt.slideshare.net/SohelRana53/mechanism-of-plant-nutrient-absorption-by-plant-root-and-plant-tissue?ref=&smtNoRedir=1).

When the roots can no longer take up water sufficiently rapidly to remain turgid, transpiration ceases and the plant wilts. The soil is said to be at the permanent wilting point. In a given soil all ordinary plants wilt permanently at the same moisture content (Tejero & Zuazo, 2018). Field capacity and permanent wilting points are the two important water-retaining capacities of soils as far as irrigation is concerned.

3.7.4. Relationships Between Soil Moisture and Plant Growth

Research work in Israel and the United States has shown that field crops grow best when the available moisture in the soil is kept at a low suction.

According to Ali (2011), when moisture in the soil is at a low suction it means that the soil has a high percentage of moisture and therefore cannot absorb any more.

As the amount of moisture is depleted, the tension increases and eventually the growth and yields of crops become affected. For most crops the reduction in yield becomes significant when the available moisture is below 50% for prolonged periods.

3.7.5. Measuring Soil Moisture Content

Over the past years, investigators have employed many methods in soil moisture studies. Some of the methods employed in soil moisture studies are briefly discussed in the following subsections.

3.7.5.1. Gravimetric Method

Soil samples secured from various depths and locations in the field is the simplest and most widely used and probably the best method for measuring soil moisture. The soil samples are placed in an airtight container which may be made of aluminum, metal or glass and sealed immediately to prevent the loss of moisture. In the laboratory, the moisture samples are weighed, dried to constant heat in an oven at 105°–110°C and reweighed (Figure 3.11).

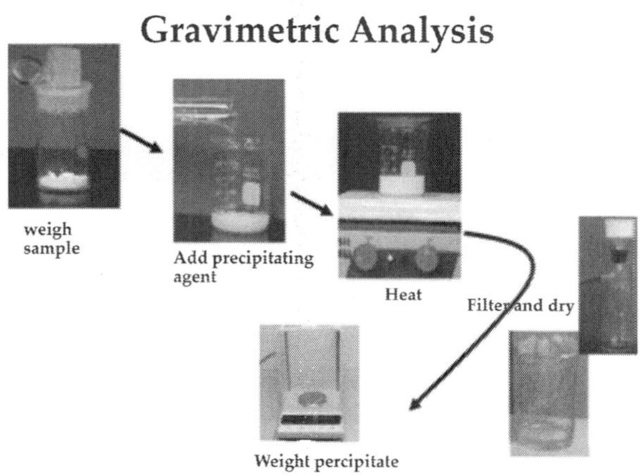

Figure 3.11: Steps used in analyzing moisture content using gravimetric method (*Source:* https://www.slideshare.net/sureshselvaraj108/gravimetry-46720672).

The difference in weight is due to loss in water and is divided by the dry weight of the soil to give the percent of moisture on a dry weight basis.

Gravimetric method of soil moisture determination is to date the most accurate. However, it suffers the disadvantage of being applicable where labor is cheap.

3.7.5.2. Electrical Resistance Blocks

Ali (2011) stated that "soil moisture measurements depend upon determination of the electrical resistance between two electrodes imbedded in a block of gypsum which is in equilibrium with the soil."

Porous absorbent blocks usually made of plaster of gypsum with the two electrode embedded are buried in the soil. Insulated wire leads connect the electrodes to a specially designed meter, which may be calibrated to read directly the percent of available moisture. These blocks are made in several sizes and shapes and are most sensitive in the lower ranges of available soil moisture (Tejero & Zuazo, 2018). They should not be used in a very wet or poorly drained soil. Their sensitive range being in the lower 25% of available range usually provides ample warning that the moisture supply is low before the soil reaches the permanent wilting percentage (Figure 3.12).

Figure 3.12: Practical application of electrical resistance blocks in soil moisture content determination (*Source:* https://geomor.com.pl/produkt/probniki-roztworu-glebowego-sps200/).

Gypsum blocks are not sensitive to moderate concentrations of salt in the soil solution, but a high concentration affects their readings and causes a deterioration of the blocks.

Other resistance blocks using nylon or fiberglass have been developed, but are sensitive to changes of salt concentrations in the soil solution and could not be recommended for soils with moderate to high salt content (Tejero & Zuazo, 2018).

Calibration for most blocks may not remain constant after they have been in the soil for some time. The blocks are placed in holes where the soil has been disturbed and roots must penetrate the region of disturbed soil before a meaningful reading can be taken. To measure the amount of soil moisture quantitatively, all those devices must be calibrated by testing them in soil of known moisture content (Tejero & Zuazo, 2018). This requires the test made for different soils occurring in the field. Nevertheless, the instruments will show that changes in soils moisture are taking place, and may be used as guide for irrigation.

3.7.5.3. Neutron Scattering Method

The neutron scattering is a recently developed method for measuring water content on a volume basis. According to Doneen (1971), it is satisfactory for measuring the water in a profile but inaccurate near the soil surface or in indicating water content in any specific narrow depth interval. These are distinct disadvantages of the method. However, measuring the soil moisture during the season at one location without disturbing the soil is an advantage in eliminating soil variation and in addition the moisture content is known immediately.

The equipment is costly and the user risks radiation hazards. At the present time, the neutron probe is still a research tool and not applicable to routine field moisture determination.

3.7.5.4. Tensiometer

Tensiometers give accurate and reproducible measurements provided they are always at the same temperature and are in both thermal and moisture equilibrium with the soil. They are calibrated to give the soil moisture tension or negative pressure (Doneen, 1971) (Figure 3.13).

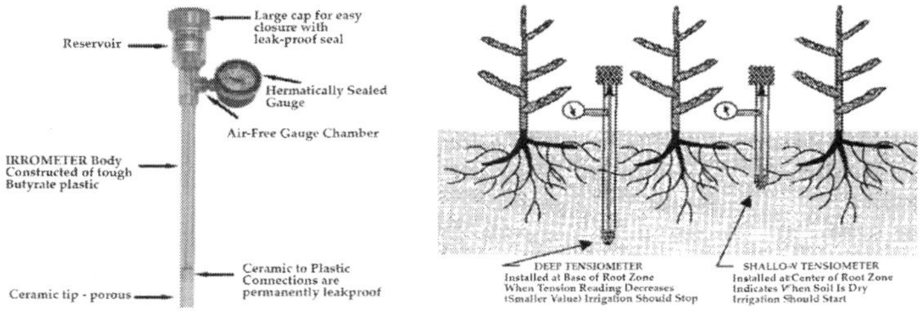

Figure 3.13: Illustrating the design and application of tensiometer in the determination of soil moisture content (*Source:* http://blog-yard-garden-news.extension.umn.edu/2016/08/useful-tools-to-determine-soil-moisture.html).

Tensiometers are limited to the higher moisture content as the instrument is restricted to less than 1 atmosphere which for fine-texture soils is less than half the available range and about three fourths the ranges for sandy soil (Doneen, 1971). The tensiometer is not recommended in estimating the soil water content unless crude estimates are adequate, then each soil should be calibrated separately. Tensiometers consist of a porous cup embedded in the soil in the zone of major root activity, with a tube connected to a vacuum gage or mercury manometer. The cup and the tube are filled with boiled water to eliminate air. In the field, considerable skill is required to maintain those instruments in a reliable operating condition (Doneen, 1971).

3.7.6. Water Response Functions of Crops

The yield of a particular crop depends upon the average growth of the plant over the whole length of its growing season. In other words, the average growth of a plant in any period is a function of the average level of available soil moisture in that period.

Irrigation experiments at Davis, California, show that yields of shallow-rooted crops may be increased several hundred percent while deep-rooted crops may be increased from a few percent up to 50%.

The effect of irrigation on yield is directly related to the amount of water held in the soil reservoir until maximum yield is produced. Additional water will not increase the yield. The Institute of Agricultural Research, Hindu University, Varanasi, India, on conducting experiments on water requirements of crops, found that average yields of different strains of cotton and sugar cane when plotted against the amount of water they consumed, gave a linear proportion between the amount of water consumed and the total produce, provided additional manure was used (Ali, 2011).

Research shows that crop yields increase with irrigation water supply within a fixed range. Higher crop response to irrigation is realized with optimum application of other inputs like fertilizers and manure. Organic manures for example modify the soil structure so that it can hold more water and air for the extra benefit of a crop. Results of experiments performed in Madras, India, indicate that the use of fertilizers increased the yield of cotton by 34% over that without application of fertilizer or irrigation. Irrigation alone gave a 37% increase, and when both fertilizer and irrigation were applied, the yield increased by 114%, which is over 50% higher than the 71% cumulative effect of the fertilizers and irrigation taken separately (Dutta, 2011).

3.7.7. Frequency and Rate of Irrigation

Under practically all types of climatic conditions there are certain periods during the year when soil moisture becomes a limiting factor for plant growth even in areas of high rainfall. Tejero & Zuazo (2018) quoted some figures indicating that in spite of a rainfall of 152 inches per year supplemental irrigation is beneficial in the period February to April although an average of 20 inches of rain is received during those three months.

Similarly, Kleinschmidt (1967) had to include an irrigation treatment as the "periods of moisture stress are common at Lawes (Australia) even during the wet season." In the tropics wet and dry seasons are in most places definite so that irrigation can be planned beforehand.

The question of when to irrigate (and when not to irrigate or to stop irrigating) is best answered when the total requirement of a crop is known. This in turn requires knowledge of how much of the incoming water (rain and irrigation) is lost in a specific time. The total requirement is governed by the length of the growing season, ambient temperature, method and frequency of irrigation, and depth of wetting.

On the other hand, water losses will be incurred through surface evaporation, surface run-off, transpiration from the foliage, deep percolation below the root zone, and use within the plant.

Penman (1963) has drawn attention to the fact that for a full cover of a green crop which is kept plentifully supplied with water, the rate of transpiration from the field is dictated primarily by the weather (radiation, mean air temperature, mean air humidity and mean wind speed), and scarcely at all on the nature of the vegetation. However, as soon as plants experience water stress or under drought conditions, the plants themselves become much more important in regulating the rate of water loss.

The total requirements of irrigation water will differ very considerably for crops with different seasons and periods of growth. Thus in an arid climate perennial crops will require irrigation throughout the year while, other crops will require water for only a few months (Snellen, 1996).

Penman (1963) developed equations whereby total evaporation (including transpiration) can be computed from the local weather data, which together with measurement of rainfall can provide the basis for the control of irrigation. However, where the desire is to maintain the growth of a crop by maintaining soil moisture at field capacity, as in the ease of perennial crops, guidance can be obtained by a quicker and simpler method.

Soil moisture can be measured as a resistance using stainless steel electrical resistance units. When calibrations for field capacity and wilting point are ascertained for the soil in question, meter readings taken at appropriate times will indicate whether to irrigate (high readings) or not to irrigate (low readings).

Most crops have two or three periods of maximum water requirement- one during the seedling stage, another during the pre-flowering and flowering stage, and the third during the seedling stage (Penman, 1963). Young seedlings are particularly susceptible to water tension, and growth is retarded if they are subjected to high water tension (Figure 3.14).

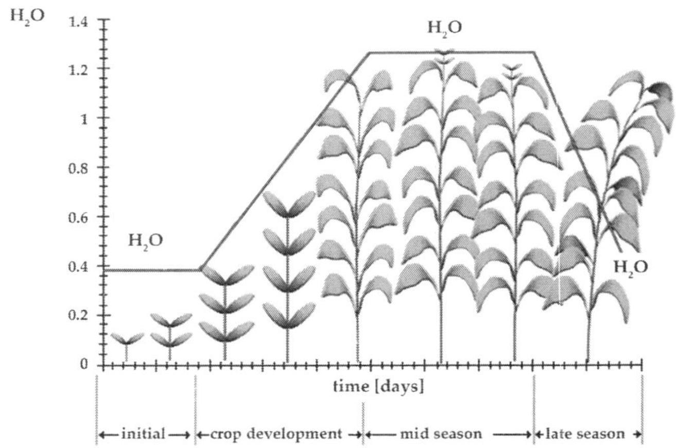

Figure 3.14: Depicting the crop stages of tomato for water management (*Source:* http://www.fao.org/land-water/databases-and-software/crop-information/tomato/en/).

For efficiency in water use, farmers need to be able to determine the proper time for irrigation. Ali (2011) gives two criteria for this purpose: "The soil becomes depleted of soil moisture which must be replenished. The soil also becomes unable to supply sufficient moisture for maximum growth and there is cessation of plant growth followed by other indications of insufficient water – a change in color of foliage and perhaps wilting."

Experienced farmers in developed areas determine the need for irrigation by sampling the soil with an auger and determining the approximate wetness by color or feel. Most soils change color between the field capacity and permanent wilting point; usually the color is darker at field capacity and becomes lighter in color as the moisture content approaches the permanent wilting point (Figure 3.15).

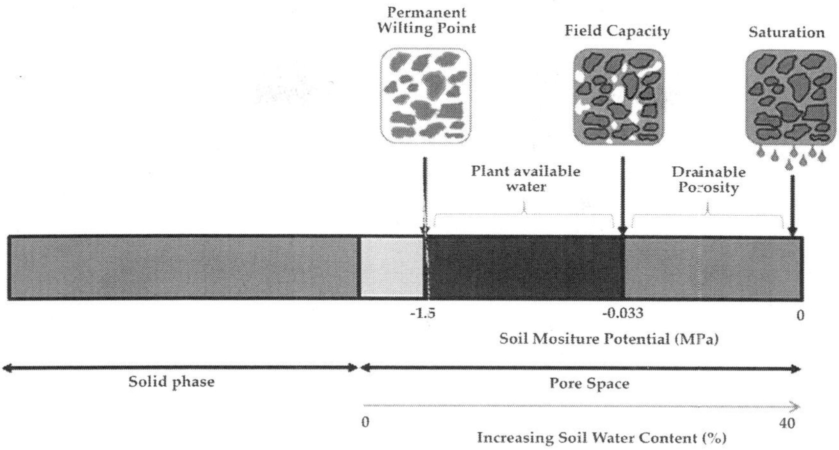

Figure 3.15: The difference in water content between field capacity and permanent wilting point (*Source:* https://www.nature.com/scitable/content/water-content-and-water-potential-at-saturation-59719594).

The color of crops becomes dark green, often almost bluish or grayish, as the supply of moisture becomes insufficient.

The quantity of water that should be applied in the form of irrigation and the frequency depend on the soil type, the crop, and the weather. The soil storage capacity, the water already held in the soil, and the rate at which water is absorbed through the soil are the principal factors in determining the quantity of water that should be irrigated.

Two important characteristics of crops which influence the frequency and the rate of irrigation are:

- The depth of the root system.
- The stage of growth in the life cycle of the crop.

The rate of transpiration will vary from practically nothing at the young seedling stage during cold, cloudy weather to a maximum during hot windy weather at the time when crops are growing luxuriantly.

Each crop has a certain, rather definite, rooting-habit which will tend to follow if the soil is uniform and deep enough and equally well-moistened. On the other hand, almost any crop will develop its major root zone in the most favorable environment with regard to both nutrients and soil moisture.

3.7.8. Water Quality as Affecting Crops and Soils

Many factors are involved in any appraisal of irrigation water. Among these are:

- The total quantity of dissolved salts.
- The particular constituents and their ratios.
- The characteristics of the soils and the crops to be irrigated.
- The irrigation practices.
- The climate particularly temperature and humidity.

Many authorities agree that the suitability of water for irrigation purposes depends on the effects of its total quantity of dissolved solids on the plant and the soil. Salts may harm plant growth physically by reducing transpiration through modification of osmotic processes, or chemically by their toxic constituents (Tejero & Zuazo, 2018) (Figure 3.16).

Figure 3.16: Illustrating the effects of salt on plants' growth (*Source:* https://www.liverpool.ac.uk/~sd21/stress/salt.htm).

Salts affect soils by changing their structure, permeability, and aeration and this indirectly affects plant growth. Soils, particularly fine-textured ones, have the ability to absorb certain minerals from irrigation water. If irrigation water contains more sodium than it does calcium and magnesium, a tendency exists for this sodium to replace the calcium and magnesium already in the soil. The presence of excess sodium in the soil makes the soil less permeable, so the effect of irrigation water containing an excess of sodium is to tighten or seal-up the soil.

3.8. SOCIO-ECONOMIC ASPECTS OF IRRIGATION

3.8.1. Cost of Irrigation

The vital role of irrigation is increasing food production and as an important starting point for overall agricultural sustainability. In view of this, important role, the development of water resources is frequently undertaken by governments. If all costs are included in pricing such water for agricultural purposes, farmers frequently cannot afford to pay for the water. Water is therefore frequently subsidized in one form or another. In developing countries, the bulk of the available water is used for irrigation.

An economic evaluation of alternative uses of water shows that agriculture is far less productive in its uses of water than are other users. The productivity value of water in industry is frequently 100 times or more as great as for agriculture. In the San Juan basin in Colorado and New Mexico (USA), productivity of an acre-foot of water in irrigation was estimated at $6 to $16 as compared with $1200 to $3000 when used for industrial purposes (Kang, 2010).

However, factors other than the direct financial return per unit of water used have in the past been considered to justify the existence of some expensive water-development projects. As capital is one of the scarcest resources in a developing economy, it is essential that economic and financial evaluation and appraisal of irrigation projects be done to determine the return to the scarcest resource.

Capital and operating costs of providing irrigation and the returns that can be expected under different conditions vary widely. The several factors which influence costs include:

- The type of water supply.
- The size of the project.
- Whether the project is government-sponsored or privately sponsored since this may have a bearing on the interest rates or other financing aspects.
- The climate of the area and the type of crops grown.

3.8.1.1. The Type of Water Supply

Costs of water for irrigation are naturally dependent on the source of supply, being generally lowest when drawn by gravity from flowing streams or springs, higher when pumped from shallow wells and streams and highest with water pumped from deep wells and reservoirs created by constructing large-scale dams. Thus for example, in the USA, studies showed stream-flow cheapest, pumping from stream next, and pumping from wells highest (Kang, 2010). The generally accepted belief that irrigation from groundwater supply is much more expensive than that from gravity-flow canals has however been disputed by (Ali, 2011) (Table 3.2).

Table 3.2: Outlining the Advantages and Disadvantages of Groundwater in Irrigation Schemes

Advantages of using groundwater	Disadvantages of Using Groundwater
• It is often available at or near the point of use and consequently does not require a complex distribution network. • Although it is generally considered more expensive than direct river diversion, this method frequently is considerably cheaper than surface storage of water by dams and is usually easier to develop. • There is less fluctuation in supply than may be the case with stream flow. • Groundwater also tends to be freer from a soluble mineral load than surface water. • It is particularly suited in regions far which surface waters are adequate far irrigating only relatively small areas, for areas isolated from streams, and for providing stand-by or supplemental facilities. • In arid areas where no perennial rivers flow, the development of groundwater resources may be the only practical solution to the problems of water supply.	• Sometimes it is available only at an excessive depth or in inadequate quantities. • Occasionally it may be of poor quality because, although usually unpolluted and relatively free of sediment, it is often highly mineralized. • In same circumstances it may prove more expensive than surface water because it requires expenditure of energy for pumping while surface waters can flow by gravity or, at times, even be used to produce energy. • Most important is the fact that there is a finite amount of groundwater available in any one area, so that if extraction exceeds infiltration, the reserve of water accumulated over prior years will sooner or later become exhausted.

Source: http://slideplayer.com/slide/6244548/.

3.8.1.2. The Size of the Project

The second factor that influences the cost of irrigation water is the size of the project. Costs of water per cubic meter from reservoirs tend to be lower for large-scale projects. Thus for example studies in Madras, India, showed costs of water, based on construction of dams, ranging from 6–9 cents per cubic meter for small dams to 0.7 cents for the largest dams (Dutta, 2011).

The cost of water whether from streams, reservoirs or wells depends on the capacity of the pumps and heights of pumping, being lower for bigger capacities and shallow depths.

It is interesting to note that costs are lower for smaller pumps up to a depth of about 16 m but apparently are much higher at depths of 20 m or more. In general, a diesel pump will use 0.45 kg (or 0.53 liters) of diesel per Hp per hour when raising water from 12 m.

3.8.1.3. Government-Sponsored or Privately Sponsored

Although theoretically the cash costs should be the same regardless of who sponsors an irrigation project, government-sponsored projects have been found to be more expensive than private projects, perhaps because only governments are willing to undertake those that are costly in both total and per terms (Dutta, 2011). There appear to be substantial diseconomies of scale in government tube-well projects.

In India and Pakistan, the well, power line, and drainage all cost more per cubic meter per hour as the project is enlarged. This may be explained by the large overhead costs involved and the often poor water distribution and management problems encountered in such government projects (Figure 3.17).

Figure 3.17: Illustrating a well-managed irrigation scheme (*Source:* http://www.wikiwand.com/en/Irrigation_management).

According to a study done by Moorti & Hellor (1969) on costs and benefits of irrigation, it was found that the original investment, besides electricity transmission for State tube-well was more than 15 times the investment for a private tube-well despite the fact that the discharge of the State tube-well was only twice that of the private tube-well.

From this comparative study, it was also revealed that most private tube-wells provided greater availability and reliability of water supply than State tube-wells, which was reflected in the higher cropping pattern and cropping intensity in the private tube-well farms. It can be concluded from this study that for almost all crops, the gross returns per hectare was higher on the private tube-well farms relative to farms irrigated by the State tube-well.

3.8.1.4. The Climate of the Area and the Type of Crops Grown

The climate of an area and also the type of crops grown in an irrigation project has a bearing on the irrigation costs and benefits. In heavy rainfall areas a few acre-inches of irrigation may be enough to supplement the rainfall whereas in dry areas plant water requirement is met through full irrigation.

4
CHAPTER

CULTURAL FARM PRACTICES THAT CONTRIBUTE TOWARD AGRICULTURAL SUSTAINABILITY

CONTENTS

4.1. Introduction.. 80

4.2. Crops Support... 80

4.3. Crops Spacing 83

4.4. Crops Pruning... 84

4.5. Mixed Cropping 86

4.6. Mulching.. 90

4.7. Crops Fertilization 102

4.1. INTRODUCTION

There are several cultural practices that are normally performed by different farmers, especially small-scale ones in order to enhance the production of their crop. Some of these practices have been discussed in detail under this chapter. Some of these practices include: crops' support; crops spacing; crops pruning; mixed cropping; mulching; and crops fertilization.

4.2. CROPS SUPPORT

Other than the normal management in the farm, there are some extra practices that are important when they are performed for the crops. One of these practices is providing support to crops. Most crops such tomatoes, peas, some beans, passion fruit among others will climb up objects such as trees or fences by twining around the support. In commercial production it is therefore necessary to provide some means of support.

4.2.1. Types of Support

Plants may be trained up living supports, but the most common system of support used in commercial plantations is by use of plain strong wires strung on erected wooden posts; this being referred to as a trellis. In view of the fact that the mature fully developed plants are heavy, it is essential that they are trained on a firmly erected strong trellis (Markham, 2010).

To get the best out of climbing crops, it is advisable to train the vines on to the trellis. This has the effect of providing the maximum possible area for growth and the formation of buds. In commercial fruit plantations trellises are mainly of two types, the vertical and the horizontal trellis.

Experimental evidence has shown that a low trellis (short posts of 1.2–1.5 meters high) of either vertical or horizontal type, results in the production of laterals which tend to trail on the ground at an early stage (Markham, 2010).

For crops such as passion fruits, high trellises obtained by using 2.5–2.7 meters high posts to allow 60 cm in the ground and at least 2 meters above the ground have been considered more suitable. Such high trellises permit plenty of room for growth and expansion to bearing laterals before they trail on the ground and allow them ample light and air (Dris & Jain, 2004).

4.2.1.1. Vertical Trellis

The vertical trellis system is commonly used in Western Australia, Victoria, and in Kenya. Its main advantage is that it is simple and cheap to construct (Figure 4.1).

Figure 4.1: Illustrating vertical trellis in tomatoes production (*Source:* https://www.pinterest.com/pin/265430971764108547/?autologin=true).

In this system wires are stretched one above the other as in ordinary fences. In Western Australia, wires are mainly used, where one wire is fixed about 8 cm from the top of the post and the other one 45 to 60 cm below it. In Queensland, three wires are used; one wire is run as close to the top of the post as practicable, and a second and a third wire, are run usually at 30 and 45 cm spacing below, 35 cm being the average spacing between the wires.

Although horizontal trellis system is mainly used in South Africa, a vertical trellis with four to five wires, one above the other with an average spacing of about 30 cm between the wires is sometimes used. In Kenya, however, only one wire is used and is fixed at the top of the posts.

4.2.1.2. Horizontal Trellis

Gilkeson (2011) stated that yielding capacity of most climbing crops is closely associated with the space allotted to the plants. For this reason the horizontal trellis system is considered to be more suitable in Queensland and in South Africa, since there is a wide space for growth expansion in this system (Figure 4.2).

Figure 4.2: Horizontal trellis in tomatoes production (*Source:* https://www.indiamart.com/meeragroinfra/crop-support-net.html).

This system is also extensively used in Western Australia. In Kenya, vines trained on a horizontal trellis system showed higher yields than those trained on a vertical trellis system.

In a horizontal trellis the wires are run side by side. The distance between the units may be anything from 25–60 cm. Toensmeier (2007) stated that wide spacing between the wires has the advantage over the closer spacing, in that it permits the entry of sunlight and air between the sets of laterals. This promotes flowering and setting of fruits on the inner growth of the vines. In order to keep the wires apart on a horizontal trellis, a T-piece (cross-arms) is fixed to the top of the posts. One wire is stretched at the top of the posts while others run parallel at the ends of the cross-arms.

This trellis system assists in maintaining a more open growth, allowing dead and diseased leaves to fall to the ground freely, carrying with them any fungus spores adhering to them. Toensmeier (2007) also noticed that this system helps in the control of the brown spot fungus by allowing mature and dead leaves to fall to the ground.

A horizontal trellis with more than three wires in which many leaders are grown along the side wires and side laterals trained over outside wires, have however, been discouraged in Queensland, as it often results in a mat or shelf of vines on the top of the trellis (Toensmeier, 2007). This holds dead and diseased leaves instead of permitting them to fall to the ground.

4.3. CROPS SPACING

According to the observation made by Markham (2010), the optimum plant density and spacing naturally varies with soil, climate and economic factors. Factors which may influence decision-making on spacing are the profitability of early intercropping, the incidence of disease (which may be increased by high plant density) and the ease and economy of mechanical cultivation spraying and harvesting.

4.3.1. Soil Moisture and Disease Factors

According to Dris & Jain (2004) observations, the more fertile the land, the greater should be the distance between the plants in the row. Planting too close in fertile land has no advantage, since with time, the foliage becomes very dense and it may then be necessary to remove every alternate vine if optimum produces are be achieved (Dris & Jain, 2004). Rich soils, warm, even temperatures, with high rainfall, contribute to vigorous growth and make the wider planting distance preferable (Figure 4.3).

Figure 4.3: Demonstrating the effect of spacing in maize crop production (*Source:* https://www.pioneer.com/home/site/us/agronomy/library/corn-planting-depth-spacing/).

From a spacing experiment carried out on passion fruits at Alkmaar, South Africa, a spacing of 3 × 3 meters gave better yields than a spacing of 3 × 6 meters, differences being particularly large on poor soils without effective fertilization. At University of Rio de Janeiro, Brazil, Aurajo (1962) observed yellow passion fruit spaced at 3 × 4 m giving better yields than those spaced at 3 × 6 and 3 × 8 m.

Moisture supplies available in the soil have also been indicated as a factor in determining planting distance. Where moisture supplies are limited, wider spacing is necessary. Toensmeier (2007) noted that the root system of pea plants warrants a wide spacing even though it is shallow; this is because their aerial and subterranean development is extensive. Wide spacing has also been noted to be advantageous in minimizing the disease and pest infestations as the foliage is kept sufficiently open, to admit light and allow free circulation of air in the canopy.

4.4. CROPS PRUNING

4.4.1. Purpose of Pruning

Pruning is an operation/activity which regulates and controls growth, flowering and fruiting. Many trees, shrubs and vines as they develop towards maturity accumulate a number of dead twigs and branches which should be removed as they may hinder further development and often harbor pests and diseases (Pittenger, 2014) (Figure 4.4).

Figure 4.4: Pruning exercise in crop production (*Source:* https://www.linkedin.com/pulse/pruning-can-increase-crop-production-encourage-growth-dr-mohammad).

4.4.2. Agronomical Advantage of Pruning Crops

Pruning of crops should aim at making room for and encouraging new growth on which the crop is borne. Research shows that pruning of crops increases the production of crops and controls cropping. Pruning also stimulates the vine to healthy vigorous growth and greatly increases its productive life (Pittenger, 2014).

Seasonal pruning of crops is considered important not only to lengthen the economic life of the vines, but also to maintain yield and to some extent regulate the time of bearing. It has been observed that light pruning of vertical trellis trained plants led to a 25% increase in yield over unpruned or severely pruned plants. On the other hand, experimental evidence from South Africa has shown that unpruned vines consistently out-yielded those which had been pruned; yields and growth were lowered by about 35% by pruning.

According to Starrs & Goin (2010), laterals pruned to two or three internodes of the passion fruits from the main leader are superior in growth and gave significantly higher yields than those pruned to 4, 5, 6 or 7 internodes and unpruned vines. It was observed that the production of the number of fruiting laterals decreased as the severity of pruning increased, but the reduction of fruiting laterals led to increased fruit size. They also observed that flower production (as reflected in the harvests) was continuous; this exhausted the pruned vines and limited the production life to four years. Nun-pruned vines however, lasted for a longer period but with a poorer performance.

Pruning of unproductive laterals should thus be reasonably close to the main leader. This results in development of long productive branches (laterals). According to studies, if crops are left unpruned for a long time, the yields of the plants are usually affected.

In Victoria, it was observed that in plantations left un-pruned, both yield and quality as well as the vines suffered, and recommended that bearing plants should be pruned most often (Starrs & Goin, 2010).

4.4.3. Pathological Advantages of Pruning Crops

Pruning has been regarded as one way of preventing the spread of pests and diseases. According to Markham (2010), pruning facilitates management and sanitation of crops in the field.

Removal of unproductive and dead laterals opens up the vines' canopy to allow a free circulation of air. Although there is no evidence to show that pruning increases the quantity of yield, it is regarded necessary from the point of view of disease control and prevention of mechanical damage plants. Pruning is necessary where the disease is present. All infected laterals are removed, as the lesions on the stems are focal points for distribution of disease and, if allowed to remain, make other control measures less effective. Fungicidal or insecticidal sprays penetrate more easily in the plants' canopy after diseased and unproductive laterals have been removed, In Kenya, pruning has been discouraged in the past as it was thought that it encouraged the spread of the virus disease woodiness. Recently Starrs & Goin (2010) were of the opinion that pruning is essential for effective control of brown spot with sprays.

4.5. MIXED CROPPING

Subsistence agriculture in the developing world largely consists of intercropping systems which enable a farmer to grow more than one crop in the same field at any one time. In this system both grain and leguminous crops which comprise a large part of the diet are mostly found growing together (Figure 4.5).

Figure 4.5: Mixed farming practices (*Source:* https://dianabuja.wordpress.com/2010/10/22/mixed-cropping-a-successful-organic-strategy/).

In many developing countries, traditional agricultural systems are based on the growing of crops in mixtures. Mixed cropping has for many years been recognized as traditionally popular farming system. For most of the small farmers almost entirely on what they can grow, it is necessary that a variety of crops be grown to allow for varied diet. The farmer sees mixed cropping as a means of ensuring this diversification.

In developing world where population pressure is reducing the area of cropping land at each family's disposal, mixed cropping systems may continue for many years if they are more efficient in using land (Grigg, 1974).

Already the system has been practiced for centuries in almost all parts of the world where agriculture is possible until very recently when Western technology started to replace the system by introducing monocropping systems. The main justification of pure stand would be high level of mechanization; which is not possible in these small farms. The scarcity and high-cost of labor for weeding among the small farmers may be overcome by mixed cropping especially where maize and beans are involved.

This is probably due to the fact that beans act as smother crop with a closed canopy. By providing better ground cover mixed cropping has encouraged the benefits of soil conservation and retention of soil fertility in most parts of the country.

In Africa today at least 70% of the agricultural populations derive their diet from mixed cropping subsistence agriculture. This indicates clearly that more research efforts should be redirected in order to promote and recognize the place of mixed cropping systems in developing world without mechanized agriculture (FAO, 1992).

4.5.1. Previous Work on Mixed Cropping

Mixed cropping as a system of farming is an old traditional method used in many countries of the tropics. Literature on its various aspects in scanty but is accumulating fast. Benson (1991) pointed out possible advantages of mixed cropping among the small farmers. In the same paper, he also expressed his opposition to agriculturists who recommended that the small farmer in Africa should adopt pure stands of crops as in Europe. Some experiments with legumes and cash crops have shown some advantages in total return of the mixtures.

Munro (1958) compared two planting dates with cotton and found that if cotton planting was delayed for three weeks the yields of maize under alternate row cropping was not increased, but the yields of cotton were reduced. Evans (1960) showed from his experiments conducted in Tanzania that intercropping groundnuts with maize or sorghum generally gave increased total production per unit land area compared with pure stands of these crops. He observed that about 12–49% more land under pure crop was required to produce the yield of one acre of intercropping depending on site and season. In addition, Belshaw (1979) reported increased returns in Tanzania for intercropping castor bean with groundnuts.

Alexander & Genter (1962) observed from their work in the USA that maize can produce about 30$ more yields, when planted in alternate pairs of rows with soybeans for the area actually under maize, than when it was sown alone. They also confirmed that soybeans produced approximately the same yields as would be expected under monoculture conditions.

Similar reports were made by Joshi (1965), from cotton intercropped with groundnuts under irrigation as cited by Fisher (1972). In India, some work was done on Castor mixed cropped with groundnuts by Reddy & Rao (1965). Their recommendation was that there was an increase in cash return of as great as 62£ over that of an equivalent area of a pure stand of either crop.

Evans (1963) has studied fertilizer and manure responses in mixed crops in Tanzania. He found that maize gives large, highly significant responses to fertilizer and manure and these responses were of the same order as for pure stands. Groundnuts did not respond when intercropped even though in pure stands responses were 250 and 300 kg/ha for manure and fertilizer respectively. Webster & Wilson (1966) in their review of some previous experiments on mixed cropping in the tropics found in most of these experiments that more than one acre of pure stand was required to produce the yield of one acre of the mixtures and reached a conclusion that for the small farmer of the tropics, there was no advantage to be gained by the replacement of the traditional practice of mixed cropping.

In addition Agboola & Fayemi (1971) in their first trials on the intercropping of maize with various tropical legumes in Western Nigeria found that the production of either crop per unit area in the mixtures was lower than their respective pure stands. They found that legume yield was significantly suppressed by maize shade. Andrew (1972) also showed from his experimental results in Northern Nigeria that there was a total yield advantage of up to 80$ from mixtures of sorghum with millet or cowpeas over the sole crop of sorghum. Willey & Osiru (1972) found from their studies on mixtures of maize and beans conducted at Kabanyolo, Uganda that yields of the mixtures were up to 38$ higher than could be achieved by growing the crops separately. Willey & Osiru (1972) talked of 55$ higher yields being achieved in mixtures of dwarf sorghum and beans over these same crops in pure stands. Norman (1972) has indicated from his studies carried out in Northern Nigeria that the profitability of crop mixture was about 60$ higher than from sole crops.

He also found that labor was used more efficiently. He based his data collection on annual crop varieties grown locally under indigenous technological conditions. In addition, Fisher (1972) in his review of the productivity of mixed cropping in the tropics expressed his strong support for the system and concluded that where hand tools are used for weed control and pesticide use is rare, there is no justification for imposing the biologically inferior European methods on the small farmer.

Enyi (1973) reported from his work of intercropping maize or sorghum with cowpeas, pigeon peas or beans in Tanzania that the grain yields of the cereal crops were reduced. He also found that intercropping maize with either beans or cowpeas decreased total grain yields of maize and legume, per hectare. But intercropping sorghum with pigeon peas increased total grain yield per hectare. In his results, he showed that the grain yields of the legume crops was significantly higher in sorghum than in the maize plots. His work confirmed the earlier studies that the combined yields from the intercrop were higher than the yields of either crop as monoculture.

Although the system of mixed cropping has for many years assured the small farmers' subsistence, it was only started by research workers about 20 years ago. The object of the small farmer is primarily to produce a reliable and varied food supply and this requirement leads him to favor mixed cropping. Leakey (1934) in describing a system in one part of Kenya mentioned that it was quite traditional and it was not easy to discourage farmers from it. He also observed that it seemed to have some advantages for the small farmers.

Generally, it has been observed in the past that many research workers have not shown interest in mixed cropping. A few such as Fright (1955), Vickers (1960), and Grimes (1963) were apparently the first workers to carry out research on mixed cropping. The reluctance in the part of other workers may be due to the variable crop combinations in use end the association of the system with traditional hand farming methods at low levels of production. This has led some of the research workers to express their doubts whether any positive benefits of mixed cropping could be exploited at more advanced level in the country.

However some experiments on mixed cropping of maize with beans in particular have been carried out in recent years by research workers based at different Agricultural Research Station and Agricultural Institutions. The results and details of their work are in annual reports and other unpublished documents.

4.6. MULCHING

4.6.1. Definition of Mulching

According to Guyer (2018), the term mulching can be defined as "the application of a covering layer of material to the soil surface." There are several material used in mulching. Some of these materials include:

- Crop residues.
- Leaves.
- Clippings.
- Bark.
- Paper.
- Plastic films.
- Petroleum products.
- Polyethylene sheets.
- Gravel.

Examples of mulches that have been used in agriculture

- The dust mulch, created by finely pulverizing the upper layer of soil.
- Plant residues, created by allowing plant materials to stay in a layer over the soil surface.
- Stubble mulch, created by permitting residues of small grain crops to remain standing in the field.
- Straw mulch, created by combines blowing out small grain-straw over fields at harvest time.

4.6.2. Effect of Mulches on Soil Temperature

Mulches applied to the surface of a soil affect the amount of heat received and the way it is dissipated. Mulches of dead vegetation, to some extent immobilize the air within the mulch and because still air has a very low thermal conductivity, heat is only slowly conducted from the surface of the mulch to the soil surface. The physical characteristics of mulch such as color, roughness and manner of application are important factors in determining the way soil temperatures will be affected.

4.6.2.1. Organic Mulches

The general effect of organic mulches is to increase soil temperatures in cold seasons and to decrease it in hot seasons. Mulches have been reported to influence soil temperatures in the following ways:

- The mulch acts as an insulating layer on the soil surface, reducing the amount of heat that enters the soil.
- A lesser fraction- of incident radiant energy is converted into heat at the surface if the reflection coefficient, of the mulch exceeds that of the un-mulched soil.
- Evaporation is reduced, which means a smaller fraction of the total heat generated at the surface is used as latent heat of vaporization, such that sensible heat constitute the larger fraction as compared with the un-mulched soil.

In USA, maize mulched with oat straw, showed that for the early part of the season mulch decreased the weekly average maximum soil temperature at 10 cm depth. The differences between mulched and unmulched treatments were not as marked in the minimum soil temperatures as for the maximum (Campbell, 1991) (Figure 4.6).

Figure 4.6: Organic mulching (*Source:* https://bonnieorganics.com/all-about-organic-mulch/).

Research has shown that cotton pods as mulch reduce the mean soil temperature at 7.5 cm depth by 1 to 2°C below that of other mulches such as black asphalt mulch. In a coffee nursery trial in Kenya, grass mulch restricted the rise of soil temperature during the day by 5°C, but the soil temperature under petroleum mulch was comparable to that under bare soil (Omosa, 1998). Lower soil temperatures under straw mulch were shown to be a major factor in reducing maize growth in the United States of America. In a straw mulch trial on the growth Alfalfa, the maximum soil temperatures at 2.5 cm depth were shown to be 9°C below the maximum for bare soil treatment (Evenson and Rumbangh, 1972). The usual effect of straw mulch is to lower soil temperatures, during the summer and increase it during the winter.

The thermal conductivity of mulch is usually much lower than that of the soil, the heat gain or lose is therefore less under mulch.

The annual mean range between maximum and minimum soil temperature was reported to be reduced by one half under sugar cane trash mulch as compared to paper mulch, the bare soil temperature range being lower than that under paper mulch (Magistud 1935).

Straw mulch at the rate of eight tons per acre reduced soil temperatures at 2.5 cm depth by as much as 17.7°C during the day (McCalla and Duley, 1946). According to Parker & Larson (1962) observations, they reported 1° to 2°C lower soil temperatures under straw mulch than under bare soil, differences between straw mulched soil and bare soil minimum soil temperatures were reported to be small, the mulched soil was slightly warmer while the soil was cooling and slightly cooler than bare soil when the soil was warming.

4.6.2.2. Artificial Mulches

Studies have shown that the temperature of the soil is greatly influenced by the color of plastic mulches. For instance, white or reflective plastics decrease or have no effect on soil temperature. On the other hand, all clear plastics have been shown to consistently result in higher soil temperatures than bare soil. Results with black plastics are variable some reports indicate higher and others lower soil temperatures, while others report little effect (Guyer, 2018) (Figure 4.7).

Figure 4.7: Plastic mulching (*Source:* http://stevia.blogspot.com/2014/01/important-of-plastic-mulch-in-stevia_17.html).

According to Omosa (1998) observation while working on maize in Kenya showed that clear polyethylene mulch was effective in creating higher soil temperatures than bare soil. Hopen (1965) recorded higher soil temperatures under clear polyethylene mulch while the black mulch was intermediate and bare soil had the lowest soil temperatures. Clear plastic was shown to increase soil temperature as compared to bare soil at 2.5 cm depth. In an experiment carried out to determine the influence of gravel, straw, black painted gravel, aluminum painted gravel and plastic mulches on soil temperatures, soil temperatures were highest under clear plastic, followed by bare soil, black painted gravel, aluminum painted gravel and straw covered treatments respectively (Hanks, Bowers, & Bark, 1961).

Block granular mulch (Coke) showed higher soil temperatures at 15 cm depth than bare soil during the day. The pre-sunrise readings did not show detectable temperature differences between mulch and bare soil (Hanks, Bowers, & Bark, 1961).

Black asphalt mulch was reported not to affect soil temperatures. Polyethylene coated paper mulches have been reported to increase soil temperatures. Paper mulch was shown to increase soil temperatures. A stone mulch increased soil temperatures compared to bare soil, which was suggested to be a factor in production of high quality grapes on stony elopes (Parker and Larson, 1962).

4.6.3. Effects of Mulches on Soil Moisture

According to the observation made by Campbell (1991), mulches affect soil moisture through increased in filtration, run-off control, reduced evaporation and weed control. Mulches have been reported to conserve soil moisture and thus resulting in improved water-plant relations.

4.6.3.1. Infiltration Rate

Vegetable mulches have been shown to improve the infiltration of rainwater into the soil, which is a major factor in the conservation of soil moisture by these mulches.

Guyer (2018) showed that the application of wheat straw mulch at the rate 1, 2 and 4 ton/acre maintained very high infiltration rates resulting in essentially no erosion.

Clean weeding caused on average 15% reduction in infiltration rate during very heavy rainstorms compared with minimum weeding or when grass mulch was incorporated into the soil during cultivation (Pereira, Feddes, & Gilley, 2013).

High infiltration rates can be maintained by keeping a continuous stubble cover on the ground. Mulches not only increase infiltration rates of rainwater and minimize its direct impact on the soil but also decrease the run-off velocity

4.6.3.2. Run-off

Mulches are generally accepted as erosion control measure. This is mainly due to their effect on run-off control. When run-off is checked the rainwater is accepted into the soil and therefore higher soil moisture contents will result. Guyer (2018) reported that surface covers of straw and gravel mulch increased water intake by reduced run-off. The protective action of a surface cover is the interception end absorption of rain drop impact which prevents, surface sealing and preserves the structure of the immediate soil surface, run-off is thus reduced by mulching and this aids to conserve soil moisture, water lost by run-off was shown to be reduced greatly by straw mulch at the rate of 8 tons per hectares (Lattanzi & Baumgardner, 1974)

4.6.3.3. Evaporation

Mulches could theoretically reduce soil water losses by reducing soil temperature, impeding vapor diffusion, acting as periodic focal points for temporary vapor condensation and absorption into the mulch material, and by reducing wind velocity at the soil interface. The reduction of soil temperatures by mulches of plant materials has been shown to be the major factor involved in the evaporation reduction process (Campbell, 1991) (Figure 4.8).

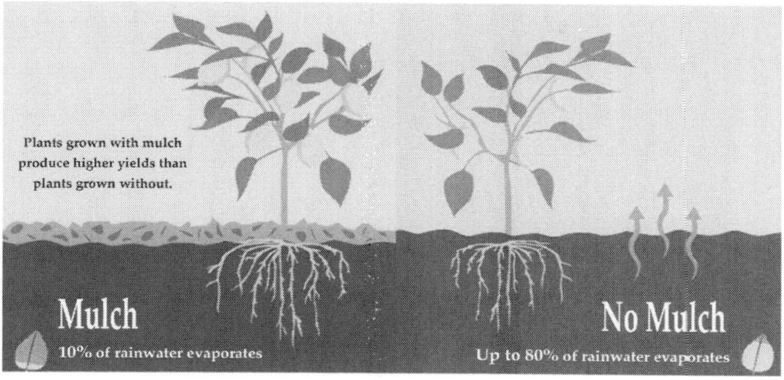

Figure 4.8: Illustrating the effect of mulching on water retention in the soil (*Source:* http://www.urbanorganicgardener.com/2016/02/mulch-much-the-benefits-of-using-mulch-in-your-garden/).

Reduced evaporation due to mulches has been reported to be one of the factors that enable mulches to conserve soil moisture. The low soil temperatures under cotton pod mulch can explain the reduced evaporation under this mulch. A stone-mulch was shown to reduce evaporation from the soil surface (Hanks, Bowers, & Bark, 1961). Straw and gravel mulches were shown to reduce evaporation thus allowing the storage of more soil moisture (Adams, 1966).

Pereira, Feddes, & Gilley (2013) reported that although evaporation reduction is a factor in improved soil moisture conditions under grass mulches, the increased infiltration rates secured by the mulch is greater than the subsequent reduction of evaporation.

Hanks, Bowers, & Bark (1961) reported that mulch acts as a barrier to evaporative losses of water to the atmosphere. A laboratory study investigating the comparative effectiveness of a dust, ground maize cob and gravel mulch in inhibiting evaporation of soil water indicated that gravel mulch was more effective than dust or ground maize cob mulch, the dust mulch proving to be the least effective. Laboratory and field experiment with granular mulch (Coke) showed that under this mulch evaporation suppression from wet sand was proportional to the relative area covered by the mulch.

4.6.3.4. Weed Control

It is generally accepted that mulches control weeds. When weeds are controlled, the water that could be lost through transpiration by the weeds is conserved in the soil for crop use (Figure 4.9).

Figure 4.9: Both figure (a) and (b) illustrate delayed weed emergence in a mulched piece of land. The mulches provide favorable conditions for crop growth (*Source:* http://articles.extension.org/pages/65025/organic-mulching-materials-for-weed-management).

4.6.4. Effect of Mulches on Soil Nutrients and Fertility

Mulches of plant materials have the effect of adding organic matter together with other plant nutrients to the soil. The amount of nutrient element added to soil will depend on the type of mulching material. This is because different materials may vary greatly in their mineral contents.

Figure 4.10: Effect of mulching on soil fertility (*Source:* https://www.sxm.com.au/news/mulching-improves-soil-structure/).

Robinson & Good (1965) working with grass mulch on coffee reported an increased organic carbon, nitrogen and reduced acidity due to mulching. The level of exchangeable potassium and phosphorus were greatly increased. Despite the reduced acidity, the exchangeable calcium and manganese were reduced. The exchangeable magnesium level was increased in the top soil but due to leaching in the greater sampling depth the magnesium levels was lowered.

The observation made by Sanders (1953) revealed that banana trash mulch increased the organic matter of the soil considerably and increased coffee yields in Tanzania. On tea soils in Kenya, mulching with grass resulted in increased percentage carbon and nitrogen in the soil. Soil phosphorus was also increased mainly in the first 2 cm of soil, so were calcium and potassium levels (Anamosa, 1990).

In Kenya, grass mulches have been shown to have strong tendency to induce magnesium deficiency by applying an excess of antagonizing potassium. Iron chloride in coffee is frequently associated with magnesium deficiency for the same reason. In coffee, potassium deficiency occurs chiefly on the older leaves.

Coconut husks have been shown to supply plenty of potassium in the soil due to their high nutrient content of 20–25% potassium oxide in the ash (Brunim, 1966).

Polyethylene coated paper mulch was shown to result in higher nitrate nitrogen levels in the soil than bare soil. The higher soil temperatures and moisture levels together with better aeration contributed to nitrogen-mineralization. In addition, to increased mineralization, the higher nitrogen levels were also due in part to reduction in leaching of soluble nitrogen (Liptay and Ticssen, 1970). In an experiment done on pineapples using straw arid paper mulches, it was shown that because straw had a high carbon: nitrogen ratio the material caused a reduction in available nitrogen in the soil. Nitrates do not accumulate in the soil until the carbon: nitrogen ratio has been lowered (Magistad 1935). Sawdust either used as mulch or mixed with soil was shown to decrease the nitrate content of the soil (White19591).

Mulches conserve the humus already in the soil by preventing overheating of the top layer and hence preventing oxidation (McMaster and Polly, 1957).

4.6.5. Effects of Mulches on Soil Structure

Mulches, in general, improve or maintain soil structure. This is brought about by the effect of mulches on breaking the force of rain droplets so preventing soil compaction, protection against exposure to alternate rain and sun also checks crusting of the soil. Organic mulches add organic matter to the soil which brings about aggregation of the soil particles (Guyer, 2018).

4.6.6. Effects of Mulches on Soil Biological Regime

Mulches modify the composition of the soil fauna and microorganisms. Grass mulch has been reported to modify the species composition of small arthropods in the soil, by increasing the participation of plant decomposing collembola and oribaticles at the expense of other species (Holler-land, 1950), Covering the soil with polyethylene mulches hindered aeration of the profile and activity of aerobic micro-organism. The populations of anaerobes are reported to increase especially in the lower layers.

Figure 4.11: High humus content enhances the population of microorganisms in the soil (*Source:* https://www.greenharvest.com.au/GreenGardenNotes/OrganicSoilImprovement.html).

Plant residues used as mulch increased the total number of microorganisms, the number of azotobacter and the intensity of nitrification in a clay soil. Peat, saw dust, and tar-paper had a negative effect which may be due to the presence of toxic substances. In another experiment, sow-dust mulch was shown to increase earthworm and nematode populations in the soil more than wood chips or peat mulches (McCalla and Duley, 1946).

4.6.7. Effect of Mulches on Soil Erosion Control

Erosion control is one of the important beneficial effects of mulches. Lattanzi & Baumgardner (1974) showed that two tons of grain straw per acre provided adequate protection to newly prepared and seeded 2.1 back slopes, when subjected to one-year frequency storms 1.3 inches of rain in 30 min. In another experiment, mulches reduced erosion by 11.7 ton/acre on 10–12% slopes. Reduction of 40% erosion has been achieved by protecting the soil with mulches (Lattanzi & Baumgardner, 1974).

The amount of surface run-off and soil erosion on a field of young tea with a ten percent slope were measured under four soil management treatments (manual weeding, herbicide weeding, oats planted between rows of tea and grass mulch), for a period of three years. The study was made in Kenya on a typical tea soil derived from a massive sheet flow of phonolite lava. The run-off and soil erosion were greatest in the absence of erosion control measures (manual weeding and herbicide weeding). Mulching gave the lowest run-off and least erosion. In general, the amounts of run-off and soil erosion were both greatest in the first year when the ground cover provided by the tea canopy was between 1 and 30%, but they were reduced to very small amounts in the third year when the ground cover was more than 60%.

The effect on run-off losses of mulching maize grown on various slopes as investigated in Nigeria, showed that mulching was as good as a forest fallow in controlling soil erosion (Adams, 1966). Surface covers of straw and gravel mulch increased water intake and essentially eliminated erosion. Stone mulch was shown to greatly decrease soil erosion. Gravel was shown to adequately control erosion on smooth, bare sandy loam soil.

4.6.8. Effect of Mulches on Plants' Growth

Crop responses to mulches come as a result of the effects of mulches on soil water, temperature, erosion, and weed control together with soil nutrients, soil structure and other effects of mulches that affect the plant growing environment.

Mulches can aid germination, emergence and seedling growth which is one of the most critical periods in the life cycle of a plant. This is achieved by moderating or improving the soil and aerial environment to which the seeds and seedlings are subjected (Guyer, 2018).

Higher soil water contents coupled with improved temperature conditions seem to be the factors involved in the beneficial effects of mulches on germination end seedling growth. Mulches generally increase soil moisture content in the zone where seedlings are normally placed.

In field and greenhouse experiments involving the application of straw or polyethylene materials as mulch showed that the rate of emergence, growth and earliness of maturity of maize was increased. In Washington USA a region where inadequate water in the seedbed at planting time is a limiting factor to early establishment of winter, wheat mulch conserved enough moisture to allow early establishment (Papendick, Lindstrom, & Cochran, 1973).

Under conditions where crusting was a factor, stabilized vermiculite and coke placed over the seeds resulted in better emergence of sugar beet than occurred with seeds placed under bare soil (Ririe & Hills, 1970).

According to Van (1959), it has been noted that low soil temperatures have been reported to be a primary causative factor in the reduced growth of maize in some parts of United States.

According to Campbell (1991) clear polyethylene mulch have been noted to induce earlier start of vegetative cycle and reduces the total growth cycle of maize by about one month.

Root development and response of sugar cane to fertilizers can be improved by mulching on saline-alkali subsoil in Barbados. According to the observation made by Evenson & Rumbaugh (1972), it was noted that straw mulch has the ability to increase alfa-alfa growth by as much as 13 cm the greatest growth difference appearing late in the season.

In a shaded coffee nursery bed in Kenya neither grass nor did petroleum mulch affect seedling growth, although small differences in soil temperatures was noted. In Tanzania mulching was shown to increase stem weight, yields and top/root ratio of coffee seedlings (Omosa, 1998). Mulching coffee with grass, there is a tendency for the feeder roots to grow into the top layers of the soil (Hanks, Bowers, & Bark, 1961). Grass mulches affect the mineral distribution of certain elements in the coffee plant.

4.6.9. Effects of Mulches on Crop Yield

Several reports have indicated higher crop yields when mulches are applied than when are not. Higher maize yields have been reported due to mulching. Cotton yields have been increased by mulching not only due to increase in soil moisture but also the effects of decreasing toxic concentrations of aluminum and manganese in the soil.

A study done by Baxter (1970) showed that peach and apple trees grew better and carried heavier fruit crops when straw mulch was used as compared to bare soil. Blundell (1954) also reported an increase in production of sugar cane after mulching with sugar cans trash. Yield increases on red currants, strawberry, eggplant, cabbage, apricot, grain sorghum, and potatoes have also been reported due to mulching (Guyer, 2018).

According to a study done in Ghana by Acquaye & Smith (1965) shown that cocoa production gave higher yields when the ground was covered with grass mulch during establishment than when the soil was left bare, or the weeds allowed to regenerate and cutlassed at regular intervals. Mulch was superior to the other treatments and the superiority persisted for three years after the application of mulch ceased.

In Malawi, Foster (1964) shown that coffee yields respond favorably to

Napier grass mulch in the absence of shade. Furthermore, coffee production in Kenya has been shown to respond favorably to grass mulches. For instance, Arton (1957) reported an increment of coffee yield over mulched treatment.

4.7. CROPS FERTILIZATION

The importance of commercial fertilizers in agricultural production cannot be over-emphasized. Farmers' enthusiasm about the success of fertilizers is clearly demonstrated by the fact that application rates for the three major nutrients – Nitrogen (N), Phosphorous (P), and Potassium (K) – have increased tremendously over the past three decades.

Figure 4.12: Application of fertilizer on field crops (*Source:* http://bio1152.nicerweb.com/Locked/media/ch55/soil-nutrient.html).

In 1950/51, total worldwide use of N, P, and K was about 4.0, 2.6, and 3.3 million tons, respectively (FAO, 1952). The corresponding consumption figures for 1978/79 are 51.0, 12.5, and 20.8 (FAO, 1979).

For the relatively immobile nutrients such as P and K, recovery by crops during the growing season may be such that a sizable proportion of the nutrient is leftover for succeeding crops. This leftover portion of previously applied nutrients is commonly referred to as "residual fertilizer" or synonymously as "fertilizer carryover."

Unlike P, and K, N is relatively mobile. All forms of applied N are usually transformed to nitrates. Mobility of N is due, in part, to the transient nature of nitrates. Nitrates are readily leached in moist and/or agriculturally disturbed ecosystems.

Nitrogen may also be lost by denitrification of nitrates, to nitrous oxide and nitrogen gasses, particularly under anaerobic conditions of the soil.

According to observation of Wittwer (1980), global annual loss of N through nitrification and denitrification is estimated to be on the order of 50–60% of total soil supply. Further losses of N may be incurred by volatilization as ammonia gas.

It is not intended here to give the impression that all forms of N are readily lost from all soil types. Nitrogen in the form of ammonium is rather stable, and its conversion to nitrate is dependent on soil and weather conditions. Nitrogen carryover is, therefore, not uncommon.

The implication of the foregoing discussion is that nutrient carryover is a general problem which is by no means restricted to P and K. It must, however, be admitted that from an agricultural production point of view, the accumulation of P is not only more prevalent, but is also more important than the case of the other major nutrients. The more mobile nutrients on the other hand pose a real danger with regard to environmental pollution (Ewers, 1999).

Where carryover is significant, making recommendations to farmers purely on the basis of current applications of fertilizer not only leads to a misspecification of production costs but also to over-fertilization of the soil. The misspecification of costs is a consequence of the neglect of the contribution of nutrient carryover to current production.

Over-fertilization, on the other hand, arises when further applications raise the total nutrient supply in the soil beyond the level required to achieve maximum crop response. It may also come about when other factors of production such as technology and management impose a ceiling on crop response. Excessive use of commercial fertilizers easily translates into economic terms particularly in view of the rapidly escalating energy prices. It also leads to a waste of scarce mineral resources and unnecessary pollution of the environment (Ewers, 1999).

According to Morris (2007) observations, farmers often rely on fertilizer use recommendations issued to them by agronomists and agricultural economists. These scientists must therefore cooperate among themselves and with farmers if a meaningful solution to the fertilization problem is to be realized.

In the past, except for a few isolated cases, recommendations on fertilizer use have been derived from static optimization techniques. The basic physical relationship was the so-called production function relating crop yield and applied fertilizer. Fertilizer-output price ratios were then used to determine the optimal levels of application (Morris, 2007).

Increasing soil nutrient build-up, high and rapidly increasing energy prices, dwindling mineral reserves and the pressing issues of environmental pollution now suggest an urgent need for more comprehensive models. Such models must take into account past cropping patterns and yields, soil nutrient levels, soil types and weather variables in addition to the conventional input-output price ratios. A time dimension must be explicit in the models. Crop fertilization should, therefore, be viewed as a long-run dynamic problem. Such a conception should lead to efficient use of fertilizer while appropriately discounting and balancing future economic consequences. This obviously transcends the notion of a production function as the basic tool for deriving fertilizer recommendations.

The motivations for the present study can now be summarized as follows:

- The accumulation of the major nutrients, or excessive use of the same, in many prime agricultural soils and the need to popularize the specification and selection of comprehensive dynamic models to be employed in constructing optimal long-term fertilization strategies.
- The need for restructuring of research methodologies in order to obtain the requisite data for estimating such models.

The provision of optimal fertilizer recommendations to farmers generally entails three broad problem areas:

- Formulation and specification of the physical relationship between crop response and factors of production.
- Collection and updating of experimental and/or survey data needed for estimation and selection of an appropriate specification.
- Using the selected model for economic analysis.

It is worth emphasizing that all three-problem areas are interrelated and that they are equally important from an economic point of view.

The use of total nutrient supply is necessary whenever carryover is significant. A consistent estimate of total nutrient supply at a given time can only be constructed from consistent estimates of carryover (Ewers, 1999).

A properly specified carryover function makes the task of estimating nutrient carryover and hence that of constructing total availability a lot easier. There is no need for terminating fertilizer applications in the farm in order to measure the rate of decline of a given nutrient. Furthermore, conversion factors obtained via the carryover function, if assumed to be stable over a specified period of time, can be used in generating future estimates of total nutrient supply in a given soil-type under a particular cropping pattern. Additional historical information concerning the soil and cropping are then no longer required. All that is needed is the soil test which acts as a sensor of the stock of nutrient supply at the end of a growing season (Ewers, 1999).

5
CHAPTER

SUSTAINABILITY OF URBAN AGRICULTURE

CONTENTS

5.1. Introduction....108
5.2. A Brief History of Urban Agriculture....108
5.3. Urban Farming for Sustainable Cities....109
5.4. The Roles of Urban Agriculture....112
5.5. Key Resources for Urban Agriculture....120
5.6. Challenges in Sustainable Urban Agriculture....129
5.7. Irrigation in Urban Agriculture....134
5.8. Wastewater Reuse in Urban Agriculture....135
5.9. Solid Waste Management and Urban Agriculture....138
5.10. Factors That Determine Land Use For Urban Agriculture....140

5.1. INTRODUCTION

This chapter reviews urban agriculture in its historical context globally and in the perspective of agricultural sustainability. The main issues in urban agriculture follow under various subheadings. These include a brief history of urban agriculture; urban farming for sustainable cities; the roles of urban agriculture; key resources for urban agriculture; challenges in sustainable urban agriculture; irrigation in urban agriculture; wastewater reuse in urban agriculture; solid waste management and urban agriculture; and factors that determine land use for urban agriculture;

5.2. A BRIEF HISTORY OF URBAN AGRICULTURE

Agricultural food production in urban areas is not a novel activity. Research worldwide shows that agriculture and livestock keeping were developed in cities and not in rural settlements. As hunters and gatherers settled in what was to become cities and centers of civilization, regular production of food was indispensable for the communities to flourish. However, the limitations of the transportation and storage systems of the day, dictated that plant and animal products for urban consumption be grown within the city boundaries, or in the immediate surrounding areas (Koont, 2017).

Initial agricultural practices in urban areas started by the selective domestication of animals and growing of crops within the pre-industrial cities and a close relationship existed between a city and its hinterland. There is significant evidence that shows that urban agriculture started way back during ancient civilizations such as the Incas of South America (Rosan & Pearsall, 2017). For instance, Aztec and Maya of Mexico, the early settlements of Java and Indus Valley in Asia, and the towns of the rivers Tigris and Euphrates are some of the starters of agriculture in urban areas. There is also sufficient record-evidence that shows that where Mexico City stands today, maize was grown on artificial farmland created by dredging mud from the bottom of the lake (Pearson, 2011).

In times as recent as the nineteenth-century, one-sixth of Paris was allocated to urban gardens, boosted by horse manure from stables in the urban area. Similarly, urban fishponds have long been perfected in Asia. In Egypt, women for decades have sold butter and cheese made from milk of urban stall-fed cows in their homes (Tinker. 1999).

In developing countries such as Kenya and Tanzania, research shows that agricultural food production in urban areas are normally associated with poor households. However, just like the rest of the world, this practice has been in existence since the pre-colonial history in Kenya and the rest of African countries. Research shows that since 1899, most of the Indian immigrants residing in railway towns such as Nairobi, cultivated urban land and sold their surplus agricultural production to Europeans. Urban agriculture has expanded greatly since 1940 s. This has been attributed by rapid population growth and grim economic prospects in the urban areas (Eberlee. 1993).

The correlation became even more significant from the mid-1970 s when the economic crisis and negative effects of SAPs compelled the urban poor to seek non-market coping strategies. From these accounts, it is evident that in history, urban agriculture developed in response to food and income needs in urban areas, prompted by available resources and easily accessible markets. As the twenty-first century begins, these are the same factors documented as incentives and opportunities spurring the activity in many cities and urban areas globally.

Dissociating agriculture and food production from cities is a relatively recent but not a universal phenomenon. Many people think that agricultural food production in urban areas started in Europe during the renaissance period, and was later spread to the colonies (IDRC, 1999). Perhaps with changing achievements in African countries, urban farming seems to have accompanied urbanization by the colonialists. However, by the end of the twentieth century, statistics indicate that more than 800 million people will harvest 15% of world's food supply by growing vegetables and raising livestock in cities (Fox, 2011).

5.3. URBAN FARMING FOR SUSTAINABLE CITIES

According to Faires (2016), mass urbanization will be the distinguishing feature of the twenty-first century. Urban centers suck up resources in form of food, energy and fresh air, and in return spew out waste. The non-urban areas carry the burden of producing the resources and acting as receptors for waste. However, with rapid urbanization, the non-urban areas are increasing overwhelmed by the burden, prompting the principle that cities should carry part of that load. Therefore, studies show that there can be no sustainable world without sustainable cities (Koont, 2017).

Sustainability implies that cities would be viable in the long-term, socially, economically and environmentally. Rapid growth and urbanization raises two main sustainability concerns.

- Sustainability of urban centers will be determined by the ways the urban need for food, water, shelter, and social organization are met.
- Survival in urban centers will depend on the way in which these social and environmental pressures are handled.

Consequently, changes must be made in the provision of goods and services to cities, in particular food, water, transportation of people and goods, disposal of waste, including land use (OECD, 2000). In history, cities have been identified with change and perceived as cultural engines driving civilization. It is not clear if they will continue to do so in view of challenges posed by rapid urbanization, increasing poverty and environmental threats, particularly in cities of developing countries.

Research shows that most cities in the world, especially in developing regions are experiencing change. However, this change is accompanied by substantive negative side effects. For instance, in 1988, statistics show that "more than 25% of the absolute poor people in the developing countries lived in urban areas. In 2000, it was estimated that this figure had increased to about 55%." Because of this increment of urban poverty, most of the urban administrators are increasingly unable to cope with pressure, especially on public infrastructure, food insecurity, unemployment, unsanitary living conditions and squalid shelters, in addition to environmental management problems (World Bank, 1999). Therefore, poverty and waste management are the two most intractable problems facing third world cities but they are also the linchpins of sustainable cities.

According to Jose (1989), urbanization has resulted in escalating urban poverty and desire to source for alternative sources of livelihood. These include agricultural practices such as poultry farming. The demand for livestock products by the increasing number of urban population encourages farming households to engage in livestock keeping. Due to quick growth of broilers, poultry farmers find the livestock system desirable though the scale is still low in developing countries (Figure 5.1).

Figure 5.1: A practical example of livestock keeping and crop growing in a typical urban area (*Source:* http://vault.sierraclub.org/sierra/201107/urban-farms/slide4.aspx).

Going by the suggestion given by Faires (2016), agricultural food production has been recognized as something that needs to be integrated into urban planning to allow access to information on best practices and avoid conflicts.

Towns have been places of food production since the dawn of human history whereby livestock were always closer home. According to Urban Harvest (2007), most of the migrants in nineteenth-century met most of their vegetable and animal food needs from their own production in summers and preserved food to eat during cold winters, while indigenous birds gave meat and eggs.

Growing food and keeping livestock in the cities has some advantages over rural farming. Some of these advantages include:

- To proximity to markets.
- Low transportation costs.
- Reduction in postharvest losses among others.

In times of conflict or other types of community unrest, urban agriculture keeps people fed when food supplies from rural areas are interrupted and hence need to develop urban farming technologies to tap the benefits of market access.

To enhance profitability of both crops and livestock production in urban areas, there is the need to take measures that will address problems affecting the performance of the production system. Some of the key problems are:

- Monopoly in supply in the supply chain.
- Better marketing.
- Storage facilities which would help small-scale producers to access market through bulking.

Research shows that the most affected are the small-scale producers who have no access to large-scale market outlets especially in rural areas.

According to a report by WorldWatch Institute (2012), it has been shown that that urban agriculture also plays a crucial role in building communities. For instance:

- Urban agriculture normally brings city residents together and helps to generate social interactions such as self-help groups of young people, women and vulnerable people who meet as a result of their involvement in agriculture.
- It also gives an opportunity to share information and skills.
- Urban agriculture also helps in improving individual well-being and voice which would not be heard.
- Through urban agriculture, people are able to get technical skills and market opportunities.

5.4. THE ROLES OF URBAN AGRICULTURE

5.4.1. Food Security

WHO (2006) defines the key components of food security as:

- food availability;
- food access; and
- food use.

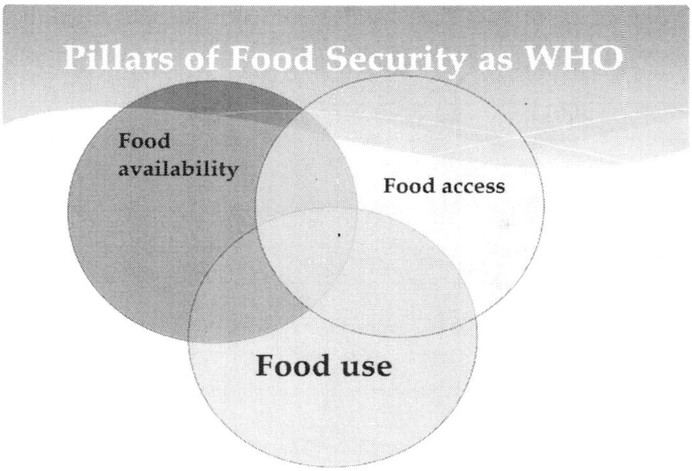

Figure 5.2: Illustrating three key pillars of food security as of WHO (*Source:* https://www.slideshare.net/rsienrose/food-security-yesterday-today-and-tomorrow).

5.4.1.1. Food Availability

This is normally achieved when safe and nutritious food is consistently available in sufficient quantities to all individuals within a country.

5.4.1.2. Food Access

This is ensured when all persons within the household have adequate resources to obtain appropriate food for a nutritious diet.

5.4.1.3. Food Use

This is normally considered in terms of quality, quantity, safety, cultural acceptability and food preferences.

It is important to note that food insecurity in urban areas is increasing, even though food and fuel continue to absorb the largest share of income in poor households. In 1990, for instance, households in nearly half of the largest urban areas in developing countries were spending 50–80% of their income on food alone (Mustafa, 1999).

Table 5.1: The Number of Households Engaged in Urban Agriculture in Some of the African Cities

Town	Country	Farming households	Survey date	Population then
11	SADC	22% (of poor)	2008	Varied
21	West Africa	20–50%	2006	Varied
Kampala	Uganda	30%	1991	774,000
Kampala	Uganda	49%	2003	1,200,000
Morogoro	Tanzania	90%	2002	228,000
Nakuru	Kenya	35%	1998	239,000
Dar-es-Salaam	Tanzania	36% (crops)	1995	2,500,000
Nairobi	Kenya	20% (crops)	1985	1,000,000
Addis Ababa	Ethiopia	17% (veg)	1983	1,400,000

Source: https://www.slideshare.net/SIANIAgri/slu-upa-ws-uppsala-diana-lee-smith-20150212.

As per the Table 5.1, in Morogoro, urban agriculture contributes about 90% of the food supply and has become an integral part of urban livelihood strategies prompted by poverty, decreased formal employment opportunities, proximity to markets, culture such as keeping livestock and non-enforcement of by laws and regulations, among others.

Urban agriculture is one of several food security options for the household. This was the principle of one of the original urban farming researches in Kenya steered by 1985 by Lee-Smith el al. (1987). The overall objective was to investigate and document the extent of subsistence food and fuel consumption in urban areas, and their significance as sources of income and employment. It consisted of 1,576 urban households in Kenya, sampled from all income groups in Nairobi, Mombasa, Kisumu, Kakamega, Kitui and Isiolo. Among them, 455 were farmers with 154 sampled in Nairobi. The method of analysis was mainly descriptive. The analysis showed that 90% of the households interviewed in Nairobi, with access to land in rural or urban areas, grew food crops.

Table 5.2: Access to Urban and Rural Land for Crop Production for Selected Towns (as % of Total Households)

Town	With access to urban land (%)	Growing food (%)	Total urban households

Isiolo	68	60	113
Kaka-mega	71	66	109
Kisumu	78	70	132
Kitui	81	79	112
Mombasa	64	55	332
Nairobi	71	65	778
All towns	71	64	1576

Source: http://www.nzdl.org/gsdlmod?e=d-00000–00---off-0hdl--00–0----0–10–0---0---0direct-10---4-------0–11--11-en-50---20-about---00–0-1–00–0--4----0–0-11–10–0utfZz-8–00&cl=CL1.3&d=HASH01776b3541c13314a9294ac0&gt=2.

Among the respondents with no access to land, 64% indicated that they would like access to urban land to grow crops. Kale, beans, maize and other vegetables were grown by 63%, 38%, 35% and 31%, respectively, of the 154 farmers interviewed in Nairobi. However, maize was the crop grown by the majority of farmers in the other five urban centers.

Since the respondents were sampled from all income groups, specific focus on the low-income households and their characteristics was lacking. Furthermore, the economic conditions which existed in the urban areas of Kenya during the time of the survey, 14 years before the current study, have worsened. For instance, the performance of the economy has continued to deteriorate despite implementation of SAPs, and the urban population living in absolute poverty has increased. Retrenchment of employees, downsizing and rationalizing of private enterprises, parastatals and public sector since the late 1990 s, has left many households in urban areas without any means of making a living.

Other urban residents in the region faced similar difficulties. In Kampala for instance, residents survived the severe economic crisis stretching from the 1970 s to the 1990 s, because a sizeable proportion of them had informal access to urban land for housing and subsistence production (Omosa, 1998). In the early 1970 s in Dar-es-Salaam, the government encouraged residents to cultivate every available space in the city in recognition of urban agriculture's contribution to food security, income, and employment. In these circumstances, livelihood strategies such as urban farming tend to become inherent because they are likely to be the next best available alternative (Trigo, 1991).

5.4.2. Generating Employment and Income

Other than food provision, urban agriculture offers employment as it generates real and fungible income. World Bank (1999) estimates that there are more than 800 million people who are engaged in urban agriculture worldwide form which 150 million are full-time employees (Mustafa, 1999).

The contribution of urban agriculture in is sometimes questioned on the basis that the poor do not own the urban land they cultivate. The view overlooks the facts that even the poor and not so, poor in the rural areas supplement their livelihood by exploiting the rural commons. For instance, income-in-kind is earned by using firewood collected from the rural commons (FAO, 1993) and from consumption of own production. The latter is accepted as part of the net revenue or profit of the farm business commonly termed as management and investment income. Use of common property resources to generate income-in-kind is however less recognized. The fact that households do not own the common property resources, pay for their use or invest in ensuring their sustainable use does not mean that they do not derive fungible income from them. On the contrary, it means that the stream of fungible income will dwindle with the exhaustion of the resource base with continued use.

The urban farmers tilling land they do not own are also tapping the urban commons to generate net returns (income in-cash and in-kind). However, exhaustion of the resource base undermines long-term benefits.

Rural and urban farmers are not the only agents exploiting the commons for livelihoods. For instance, other informal business in the urban such as small-scale retailing, street food kiosks, curios vendors or open sheds for furniture making or an assortment of repair work, are located on the urban commons particularly on the road sides and other open spaces. Demolition of these business premises does not deter owners from reconstructing them. They take the confrontation as an opportunity to challenge the government and urban authorities to give them alternative opportunities to generate livelihoods, devoid of "the obnoxious characteristics" objected to in the regularly harassed informal businesses. A similar challenge is appropriate for urban planners who object to urban farming because of its "visual untidiness."

Harare, Zimbabwe has similar dilemma visibly demonstrated by the regular street skirmishes between the municipal authorities and fruit and vegetables vendors determined to sell their commodities in undesignated areas oblivious of other urban activities (Jose, 1989).

Resource-constrained households in the rural and urban area are not the only entities tapping the services of common pool resources without explicitly acknowledging their cost. Manufacturing industries for instance, discharge effluent into rivers or smoke into the air (negative externalities to other users) incurring no direct cost. Some large-scale farmers use cultivation method that discharge fertilizer, pesticides and herbicide residues leached through the soil into rivers and lake with negative effects. They are tapping on the capacity of the environmental commons to absorb waste. Evaluation of their profitability and efficiency does not take these costs into consideration.

5.4.3. Environmental Management

According to IIED (2001), the poor in most of the urban centers do not contribute significantly towards resource degradation, except perhaps in their own neighborhoods. This is because unlike their rural counterparts, they have very little access to environmental resources. However, they are often exposed to high levels of environmental challenges such as poor quality and overcrowded housing, inadequate provision of water, sanitation, drainage, health care and garbage collection. The very large health burdens that arise from these challenges are also a major cause or contributor to urban poverty. Selective environmental improvements addressing some of these threats could therefore reduce the extent and depth of urban poverty considerably.

Urban agriculture can reduce the cost of waste collection, treatment and waste disposal, and facilitate environmental management through composting of organic waste (Faires, 2016) (Figure 5.3).

Figure 5.3: Illustrating food waste composting and vertical garden in the city (*Source:* https://www.biocycle.net/2017/09/07/composting-roundup-77/).

Alongside its beneficial products like food, money, compost and health impacts in terms of fresher, cheaper and more diverse food, especially for the poor, urban agriculture can be used to provide better sanitation and more green space in the cities for recreational value, well-being and improved air quality. In addition, urban agriculture is one of the tools of making use of open space, and managing freshwater resources more effectively by using domestic wastewater to safely irrigate selected crops.

This indicates that urban agriculture can have additional value in public health and sustainable resource management. If crop production in urban agriculture is similar to rural farming, the cost of external inputs particularly inorganic fertilizers is likely to be a constraint. Equally, continued use of inorganic fertilizer without assuring organic inputs (as manure) is detrimental to productivity in the end. Composting organic waste for urban agriculture reduces both constraints (Fox, 2011).

Characterization of urban farmers is one way of suggesting specific alternatives for urban agriculture to complement waste management objective in an urban area. The link between urban environmental improvement and poverty reduction is sometimes explained in terms of The Green Agenda and The Brown Agenda. The link creates a niche for urban agriculture in contributing to benefits of urban environmental management, enjoyed by the urban residents in general. It also gives an opportunity for the city administrators to involve the residents and the private sector in enhancing urban environmental health (Veenhuizen & Danso, 2007).

According to IIED (2001), both Green and Brown Agenda focus on environmental problems but with different priorities. The Green Agenda seeks to have human activity reach a more harmonious balance with nature for the benefit of future generations. It advocates reduction of the negative impact of urban-based production, consumption and waste generation, on natural resources and ecosystems and in turn, on world's life support systems. These are commonly perceived as the environmental degradation concerns of affluent society, because urban poor groups generally have low levels of use for renewable resources. This is not from choice or lack of need, but because the supplies are inconvenient or too expensive (IIED, 2001). For instance, the urban poor section consume less fresh water, occupy less land, consume less food per person, and have diets that are less energy and land intensive, relative to middle- or upper-income groups (Figure 5.4).

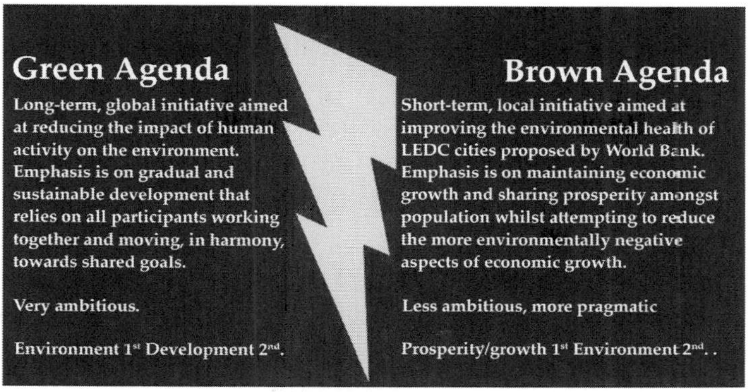

Figure 5.4: Illustrating characteristics of green and brown environmental (*Source:* http://slideplayer.com/slide/10434378/).

On the other hand, The Brown Agenda focuses on the environmental health problems and burdens typically associated with poverty and are characteristic of cities in developing countries such as, poor sanitary conditions, overcrowding, inadequate water provision, air and water pollution and accumulation of solid waste. However, the available literature shows that the use of the two terms overlaps as in the following description of Ahmedabad city in India, and Dar-es-Salaam, Tanzania.

In 1995, an empirical assessment revealed that the ambient air quality in the industrial and dynamic Ahmedabad city was a major health risk to residents. Consequently, the greening program was launched. Urban agriculture, forestry, and other greening activities were initiated as strategies for sustainable city development. The objective of the municipal administration was to reduce air pollution with jobs and income generating activities that could attract investment from the private sector and residents. The activities included greening of roadsides and traffic islands, park and garden development and maintenance, regeneration of wasteland and restoration of vacant lots. The approach asked for participation and partnerships among citizens, civic organizations and private sector. These groups however required a guaranteed of adequate time to recover their investments in crops and trees grown. Therefore, the city administrators planned, approved and set aside land with adequate tenure duration for the greening initiative (Mustafa, 1999).

In 1992, Dar-es-Salaam became one of the eight cities in Africa to adopt and apply the environmental planning and management (EPM) approach in urban planning under the Local Agenda 21 and Habitat Agenda. Through

EPM, urban agriculture was identified as one of the environmental issues to be included in the Sustainable Dar-es-Salaam Project. Working groups (formed with comprehensive representation of the wide range of interests found in the city) were set up to device strategies of ranking urban agriculture in the city agenda (Mougeot, 2006). The Urban Agriculture Working Group (UAWG) addressed broad policy and structural issues among them; insufficient planning, financial services and access to extensions from ministries, loss of agricultural and green belt land to urban expansion, land tenure issues in relation to urban agriculture, potential urban development particularly in peri-urban areas, and water use conflict between urban agriculture and other urban users. Neglecting of processing, marketing and extension needs of small urban livestock keepers and small urban farmers was underscored (Mougeot, 2006).

5.5. KEY RESOURCES FOR URBAN AGRICULTURE

5.5.1. Land

There is a common notion among many city dwellers that land appropriate for farming is not available because urban areas are commonly perceived as consisting of continuously built environment. However, the notion is inaccurate because most cities, including those in highly industrialized countries, have open spaces where food can, and is often grown (Figure 5.5).

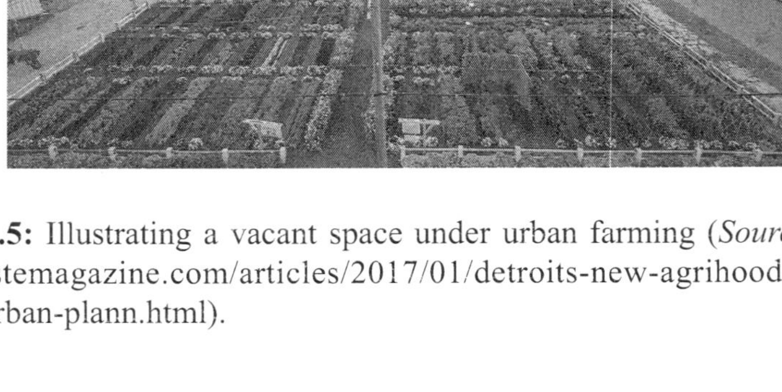

Figure 5.5: Illustrating a vacant space under urban farming (*Source:* https://www.pastemagazine.com/articles/2017/01/detroits-new-agrihood-is-the-future-of-urban-plann.html).

According to Mustafa (1999) views, most of the community gardens on empty lots are found in the urban areas of US and Europe. Moreover, the dynamic development of cities means that new spatial structures are created while others decay. Consequently, most cities often have a lot of (temporary) vacant open space that could be used for urban agriculture (Figure 5.6).

Figure 5.6: A group of youth who have converted a dumpsite into a vegetable farm near Nairobi city in Kenya (*Source:* http://www.irinnews.org/report/80150/kenya-rubbish-dump-cabbage-patch).

Studies show that only 45% of metropolitan land in Addis Ababa, Ethiopia is built up and buildings are unlikely to take all the remaining land in the immediate future. The land can be allocated to urban agriculture based on the needs of those living in there. This can be done through farmers' groups like cooperatives. Therefore, land is gained in urban areas for urban agriculture. The only thing that is lacking in most urban areas, especially in developing countries, is the official recognition of that availability making access to that land the problem. Where national and local urban planners do not recognize urban agriculture as one of the legitimate land uses, the activity faces challenges of land availability, access and usability (Rosan & Pearsall, 2017).

According to Faires (2016) views, obtainability refers to the existence of land that can be used for urban agriculture in the short-, medium-, or long-term. The first step towards obtaining urban land for sustainable agriculture involves making an inventory of available vacant land. This can be done through data collection using a combination of participatory methods and geographical information systems (GIS). Once the data is collected, it should be analyzed correctly in order to determine land suitability for agricultural use.

Accessibility to suitable and adequate land is very important element of urban agriculture sustainability. This should be within a favorable legislative framework (Mustafa, 1999). Accessibility of land refers to the opportunity for actual use of available land by households or groups, mediated through favorable legislative framework consisting of the actual administrative procedures and conflict resolution mechanisms.

Accessibility as a requirement is not unique to urban farming because rural farming also faces several challenges such as:

- inappropriate systems of land ownership;
- tenure and land transfer; and
- access and user rights.

Most of the middle-class residents access land through holding formal titles as in private ownership. They also access municipal, state or institutional land through other informal means like lobbying or even corruption. On the other hand, most of the poor access land formally or informally through squatting, sharecropping, renting, leasing, inheriting through customary mechanisms or outright purchase.

Studies have shown that the process of claiming formal access to urban land is made difficult by long and unclear procedures. Municipalities are also often reluctant to give long-term land leases (Veenhuizen & Danso, 2007). Therefore, most urban farmers usually choose alternative arrangements that are less secure. For instance, Nairobi bylaws only prohibit cultivation on public streets maintained by the city authorities. Therefore, the bulk of cultivation done by low-income households in areas that are not public streets is essentially legal but insecure because farmers have nothing they can show to back their legality claim.

The role of informal ties, networks, kinship, and information, is important as a strategy for informal access to off-plot land for urban agriculture. For instance, land is accessed through investing in social relations such as marriage, lobbying in groups with caretakers to lease land, and in some cases even occupying vacant land. As such, the very poor and those who have recently arrived in urban centers find it more difficult to access urban agriculture inputs; particularly land (Mougeot, 2006). Farmers who have cultivated for longer periods have clear advantages towards better access.

In many urban areas, urban boundaries have continued to expand to include hitherto rural agricultural land. The resulting peri-urban area is considered as the most freely available category of land for urban agriculture, because its access is determined by ownership patterns reflecting traditional and modern rules.

In urban and peri-urban Kenya for instance, a common strategy for pre-empting land grabbing is to allow a relatively low-income person informal cultivation rights pending development of the land parcel. However, it is not clear the extent to which this mode of accessing land is important to low-income households, and the extent to which different modes of informal access to land constraint profitability of urban agriculture.

Access to land for urban agriculture has five main problems and they are instructive to policy for the activity in urban area.

- Most cities and towns have not zoned land.
- Urban agriculture has not been incorporated in other social support programs that yield positive benefits as public goods. Usually these programs use community gardens to contribute to poverty reduction, facilitate local economic growth, create employment, and support urban youth initiatives. Since urban agriculture has unexploited potential to contribute in these areas, government support to urban agriculture can be viewed as an innovative way of focusing on these objectives.
- The high demand for urban land for residential, institutional, commercial and industrial use means that urban agriculture is often out-competed and ends up in marginal areas with fragile ecosystems such as wetlands, hill-slopes or in boundaries of the cities where it is tolerated until displaced by other development ventures. A possible solution is to formulate policies that do not provide land for urban agriculture based on demand and supply on the open market, but on the important role, it plays in sustaining livelihoods, particularly of the urban resource poor. However, the size of land parcels cultivated by low-income households is not known and it is not clear if it is a major constraint to returns and efficiency.
- Lack of user rights for land or brevity of their duration limits the services and resources, such as finance, that urban agriculture can attract. The disincentive arises because in most cases government or municipality owns the land used, and in other cases, it is earmarked for future development. In Accra for instance, land is leased for short duration such as one year, and renewal is not secure meaning that the urban farmers cannot plan for longer

periods. It is worth to note that recognition and in turn adequate duration of user rights was an important prerequisite for attracting participation by the private investor in the greening program of urban agriculture in Ahmedabad city, India explained in an earlier section (Marulanda, 2000).

- Access to land is limited further by contamination of soils and irrigation water. Land grabbing, corruption, and increasing land rent further aggravates the conflicts in land use, making conflict resolution mechanisms necessary. However, since land use and ownership disputes handled by the judicial system, in Kenya, for instance, are protracted, faster mechanism would be recommended for urban agriculture.

Appropriate zoning of urban land would ensure that land with potential for urban agriculture is accessed with specified user rights to encourage invest even of borrowed funds. In the process of combining urban agriculture with other community initiatives, problems faced by special groups, such as women, would be more easily identified. During zoning, concerns emanating from potential use of contaminated water and soils would be addressed with a view to pre-empting them (Rosan & Pearsall, 2017). Consumer confidence would be boosted and production targeting specific markets would be enhanced. This is a crucial component in the transformation of the activity to become a business (Figure 5.7).

Figure 5.7: Zoning of urban spaces for livestock keeping (*Source:* https://www.slideshare.net/APA-MA/planning-and-zoning-for-high-yield-urban-agriculture).

Usability of the land for urban agriculture is determined by several factors. Some of them influence accessibility as outlined in the foregoing discussion. Some of these factors include:

- Security of tenure or ownership.
- The size and previous use of the plot.
- Planned city development.
- Proposed urban farming system including potential competition from groups interested in the activity.

Water supply and its usability are closely linked to usability of land. Use of contaminated water for irrigation poses potential health risk, but on the other hand, it contains nutrients reducing the need for external inputs (Figure 5.8).

Figure 5.8: Illustrating young farmer growing arrowroot near a polluted river (although the crops benefit from nutrients in the water, this is likely to cause some health risks to the consumers) (*Source:* http://www.farmerstrend.co.ke/arrowroot-farming-in-kenya-the-high-value-crop/).

Topography, soil quality, texture, fertility and moisture content are components of environmental quality determining usability. It is not clear the extent to which variables related to usability of land such as irrigation and size of land parcel cultivated are important in influencing urban agriculture returns and efficiency among low-income households.

It is suggested that urban agriculture is caught up in a vicious cycle in that if land is available, access is denied, and if land is available and accessible, it is unusable (Fox, 2011). One option of breaking the vicious cycle is to amend the legal frameworks to recognize agriculture in the city, so that it becomes a legitimate activity able to attract resources to develop it. It is noteworthy that most by laws of urban agriculture tend to control the activity rather than facilitate and develop it. Innovative ways can be found to improve access to land and other resources and examples exist.

Research shows that the municipality of Argentina has ordinances allowing community to run urban agriculture gardens on public and private land. It has been revealed that temporary user permits are obtained from the relevant authority for public land. Furthermore, the government also encourages the private landowners to "give" their land to the municipality for a period of two years and in turn be exempted from paying property land tax. Despite the fact that cultivation of urban land pending construction type of development by low-income residents is common in most countries, especially in Africa, the available literature do not reveal any study focusing on the user rights and constraints of such urban farmers (Fox, 2011).

Most of the challenges and constraints of land availability, accessibility and usability in urban agriculture can be resolved through workable options for integrating the practice with other economic activities. Some of these options include:

- Official recognition of urban agriculture as a permanent activity in urban land use.
- Linking a wide large of functions that need policy, regulation and a specific agency to be charged with responsibility of fostering urban agriculture.
- Government facilitating urban agriculture through designating certain areas in the city
- Setting regulations on urban agriculture.
- Conducting researches, reviewing existing laws and establishing appropriate infrastructure for urban agriculture.

There are several strategies that can be practiced in order to ensure sustainable resource use, increased profitability for individual farmers from intensive use of inputs and to contribute positively towards urban environment. They included:

- Restructuring land access and land-use laws.
- Popularizing farming technique for intensive land use, floriculture and tree planting.

- Moving large livestock to peri-urban areas where fodder production is viable, composting urban waste and biogas production.
- Reclaiming previous quarry sites.
- Irrigation from wells and boreholes.
- Developing aquaculture where appropriate.
- Rehabilitating previously existing city horticultural gardens.

Additional urban agriculture strategies include:

- Information campaigns through the mass media, workshops, seminars and reading material.
- Additional economic incentives such as encouraging private land owners and institutions to allow temporary use of their land by urban agriculture, affordable livestock expense including insurance coverage, feeds, extension and veterinary services.
- Long term investment as in processing of dairy products, irrigation and water storage facilities.
- Enforcement of regulations in view of public concerns related to livestock keeping in densely built up areas, tariffs for water use in urban agriculture, urban waste management, as well as protection of water catchments and flood plains.

5.5.2. Urban Agriculture Financing

According to Pearson (2011), lack of formal recognition and security of user rights for adequate duration are some of the constraints in urban farming because they limit capital investment. For instance, lack of user rights or brevity of their duration precludes substantive planning and financing in Accra. In Dar-es-Salaam, lack of financial services has revealed as one of the main challenges most urban farmers face. In Nairobi area, Kenya, lack of inputs has been noted as the most limiting factor for urban farming, particularly for new residents.

Urban agriculture is not the only activity limited by financing. For instance, Habitat Agenda (1996) report revealed that limited access to credit is one of the many constraints to adequate shelter and income generating activities by the urban poor. The report also indicated that lack of capital and financing put at risk sustainability and returns of urban agriculture, especially by resource-poor farmers. This is mainly because most of them are usually unlikely to invest in farming technologies that conserve resources such as soil and water. This in return makes them unable to produce lucrative high-value crop and livestock products, with adequate returns for basic needs and savings to invest in urban agriculture.

According to World Bank (1999), a vicious cycle of low productivity, low savings and investment and in turn low productivity is normally attributed to lack of capital. Additionally, it has been revealed that lack of relevant information on the requirements for an effective credit scheme for urban agriculture is also another constraint that most farmers face in financing urban agriculture. This is because the little information available is based on anecdotal evidence (Figure 5.9).

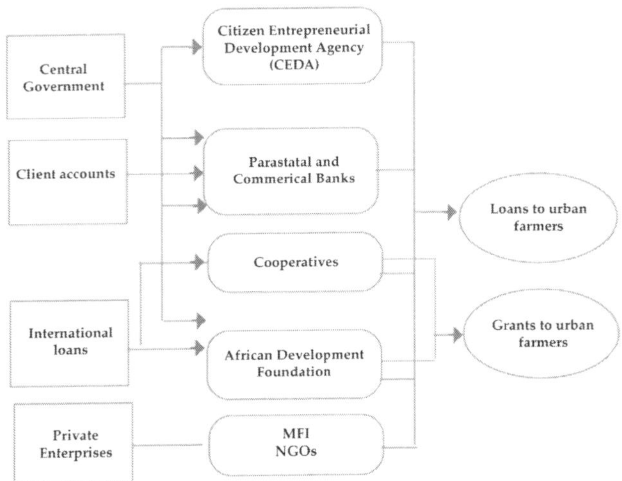

Figure 5.9: Showing major sources for urban agriculture finances (*Source:* https://prd-idrc.azureedge.net/sites/default/files/openebooks/216–3/).

Studies have shown that savings, subsidies and micro-credit are three main mechanisms best suited for urban agriculture financial support. Dubbleling and van Veenhuizen (2003) point out that if urban agriculture is viewed as part of the formal economy, it should be self-sustaining. However, if urban agriculture is understood as part of the social economy contributing to the environment and to the community, then these benefits justify external support or subsidy given to the activity. It could be done through municipal budgets for social support programs involving urban agriculture and imparting public goods, benefits. Governments also need to show their strong commitment to urban agriculture. This can be achieved through facilitating credit, financial policies and instruments, tailored to its specific needs (Veenhuizen & Danso, 2007). This was one of the recommendations in the Quito Declaration of April 2000 signed in Ecuador by the Latin American City Working Group on Urban Agriculture and Food Security, representing 40 cities in the region.

5.6. CHALLENGES IN SUSTAINABLE URBAN AGRICULTURE

5.6.1. Public Health Risk

According to Mustafa (1999), most of the municipal authorities and state officials have remained cautious and sometimes even aggressive to urban agriculture due to various risks attributed to health and environment. Most of these concerns are reasonable because the aim of public health is to promote well-being, prevent diseases and disabilities, and to enhance the quality of life. These include the physical, mental and environmental health concerns of communities and populations.

It is helpful to consider the major sources of health risk originating from urban agriculture with a view to suggesting possible ways of reducing them. One source of health risk is uptake of heavy metals from contaminated soils, water and air. The main heavy metals include lead, cadmium, chromium, zinc, copper, nickel, mercury, manganese, selenium and arsenic (Koont, 2017). There are three main causes of soil pollution from heavy metals. These include:

- Industrial refuse can contaminate fresh or wastewater used for irrigation in urban agriculture.
- Oil spills and industrial waste on previous industrial land will contaminate subsequent crops.
- Application of contaminated solid waste products to soils will contaminate crops grown.

Studies show that some heavy metals for instance, precipitate in sewerage sludge resulting in relatively high concentrations. The heavy metals accumulate in the edible parts of the crops eaten by people and animals (vanLoon & Hugar, 2005).

Nevertheless, the uptake of heavy metals varies with different type of plants and soil pH, and different amounts accumulate in different parts of the plant. This feature avails opportunities of choosing the type of crop depending on the degree and type of potential contamination. For example, it has been noted that leaves have the highest accumulation of heavy metals whiles seeds have the least. Crops like beans, peas, melons, tomatoes and pepper show very little uptake of heavy metals (Ewers, 1999).

There are several ways through which health risks associated with agricultural food production in urban areas can be solved. Some of these ways include:

- Crop selection and restriction.
- Testing of agricultural soils and irrigation water for heavy metals.
- Treatment of wastewater.
- Adoption of appropriate irrigation and management practices to control human exposure.

For instance, to reduce contamination of crops by lead and cadmium from exhaust fumes of vehicles, a minimum distance is recommended between crop fields and main roads. An alternative would be to plant boundary crops along the roads. Washing and processing of contaminated crops may effectively reduce heavy metal content such as lead in green beans, spinach, and potatoes. Some soil treatment, such as application of lime, increases the pH and in turn decreases the availability of metals (Jose, 1989).

Application of farmyard manure reduces heavy metal content of nickel, zinc and copper. Plants such as Indian grass (*Brassica juncea, L.*) when planted in hydroponics beds can be used for biological remediation of polluted soils or streams. However, more research is needed on chronic health impacts of heavy metals.

It is imperative for policies to note that the pollutants of soil, water, and air that find their way into urban agriculture products come from other processes of urban production and consumption and would exist and be harmful even without urban agriculture. In urban areas where environmental issues take center stage from the pressure exerted by urban residents, industrialists do not discharge effluents into rivers and dump sites or spew effluent laden smoke into the air unbridled.

Pollution fines and standards prevent them from exploiting the urban commons as a free receptor for waste. It has been noted that lack of regulations, inappropriate regulating mechanism and improper functioning of mechanisms established sustains some of the contaminants of urban agriculture produce. Therefore, these issues must be addressed in order to have sustainable urban agriculture. Concerns from the urban resident will increase as the environmental health risks; sources of pollution and policy failure become commonplace (Trzyna & Osborn, 1995).

Environmental health in an urban area has been noted as one of the features that tip the balance in competition for business investments. Contrary to the alternatives outlined above, the common reaction to concerns of public health is to ban the urban agriculture. However, prohibiting laws do not seem to hinder people from engaging in urban agriculture. On the contrary, practicing urban agriculture without or with limited support can increase the health risks. It is therefore imperative that health and safety concerns be systematically integrated in urban agriculture (Mougeot, 2006). The result will be reduced risks particularly for urban farmers, their families and consumers. The envisaged transformation and integration of urban agriculture with the rest of the economy is based on consumer confidence in its produce, particularly on health risk concerns. One option would be to create an agency to oversee and facilitate the urban agriculture as an industry. It would act as watchdog for compliance with laid out standards on health risks. It would also spearhead research to inform policy in this area.

There are several areas that necessity more research to inform policy. These include:

- Environmental conditions under which health problems related to urban agriculture occur in terms of characteristics such as type of agriculture, farm management practices, and characteristics of location need to be identified.
- The most vulnerable groups to those impacts and the socio-economic factors explaining their vulnerability have to be identified.
- The factors restricting urban poor from engaging in less risky agricultural and food practices need to understand.
- Resources and technical capacity available in cities to implement mitigating policy measures for health risks require and accurate assessment.

It is not clear if the farmers themselves perceive health risks as constraints in urban agriculture. It is easier to put appropriate measures in place to remove an already perceived constraint.

5.6.2. Constraints of Urban Livestock Keeping

Livestock keeping is common in low-income urban areas. However, there are a number of main constraints faced by poor urban livestock keepers likely to have impacts on urban residents and city management. These include:

- Health risks from inappropriate animal waste management will increase with increasing livestock numbers.
- Competition for highly priced piped municipal water between animals and people prompt livestock keepers to use water from alternatives sources that are usually contaminated increasing risk to health.
- Since the costs of vaccination and veterinary drugs are high, their use is sub-optimal compromising animal health, productivity and returns. Chicken, pig and duck keeping is particularly vulnerable to disease.
- Low productivity is caused by lack of feed and control over what scavenging animals eats. It is compounded by poor and low quality management practices. Poor livestock farmers have limited knowledge of appropriate livestock husbandry practices.
- They lack the skills to organize collective action to articulate their needs to urban authorities in a concerted manner.
- They also lack access to information, research and extension services. The services that could be available are not focused on their unique situation and constraints.

In particular, livestock keeping in urban and peri-urban areas is challenged by secure land tenure, limited space, capital, theft of livestock and unfavorable legal and institutional framework (Pearson, 2011).

Figure 5.10: Illustrating a group of men with their goat along Nairobi urban areas (*Source:* https://clippings.ilri.org/2015/03/31/all-things-zoonotic-an-urban-zoo-research-project-tracks-livestock-based-pathogen-flows-in-and-around-nairobi/).

Since urban livestock practice has continued and increased in importance for the poor despite these challenges, it can be concluded that its positive effects on livestock keepers outweigh the negative aspects.

5.6.3. Gender Disparity

According to Faires (2016), more women than men engage urban agriculture because they still bear the responsibility of sustenance and well-being of households and their lower education status makes it more difficult for them to find formal wage employment. However, some researchers such as Wilbers et al. (2004) have revealed that participation of men and women in urban agriculture is country specific. For example, it has been noted that most men than women have dominated open space urban farming other than traditional vegetable production in most West African cities. On the contrary, research has shown that there are more women than men practicing sustainable urban in Eastern and Southern Africa (Figure 5.11).

Figure 5.11: Illustrating a group of women engaged in urban agriculture (*Source:* http://www.farmerstrend.co.ke/how-kenyans-are-farming-in-the-city-of-nairobi/).

Because of the flexibility of urban agriculture, most people are able to respond to different gender roles. For instance:

- For women, urban agriculture enhances food security independent of market fluctuations.
- It increases women' independence because they are able to spend their money on priorities of their choice.

- women unlike men, specialize in marketing of temperate vegetable or in growing indigenous vegetables because they; require minimal capital investment, use skills and technology passed on at home through generations, allow them to cultivate less suitable pieces of land close to the residence where they can also sell and attend to chores, generate a continuous stream of income through early and multiple harvesting, and requires less strenuous work.
- Urban agriculture also allows women to earn income from a culturally acceptable activity, which in turn improves their bargaining power in intra-household conflicts.
- Women as underprivileged heads of households are able to join urban agriculture vegetable growing cooperatives since no special skills are needed.

These examples illustrate that gender is one of the key tools in unraveling the complexity of urban agriculture. Therefore, there is need to separate access to and control over productive resources such as land, credit, labor and information, from access to and control over the benefits resulting from that production. Social-cultural and economic conditions control both components. They also control the type of risks women are able and are prepared to take and determine the roles and responsibilities of men and women within the households. These interactions can only be unraveled if gender analysis becomes mainstream in research possibly (Sutherland, Wilson, & Zagata, 2014).

5.7. IRRIGATION IN URBAN AGRICULTURE

The last few decades, there has been considerable interest in the international community in developing sustainable resource management strategies for urban and peri-urban environments in developing countries. The FAO established an interdepartmental program on the subject of urban and peri-urban agriculture in 1999. Cities Feeding People program of IDRC of Canada was started in 1993. Despite these interests, the supply of quantitative information to guide technical and policy decisions have been outstripped by demand. In particular, evidence on the nature of irrigated urban agriculture was lacking (OECD, 2011).

Some of the main sources of irrigation water used by farmers in urban agriculture include: rivers and streams, raw sewerage water and piped city council water (Figure 5.12).

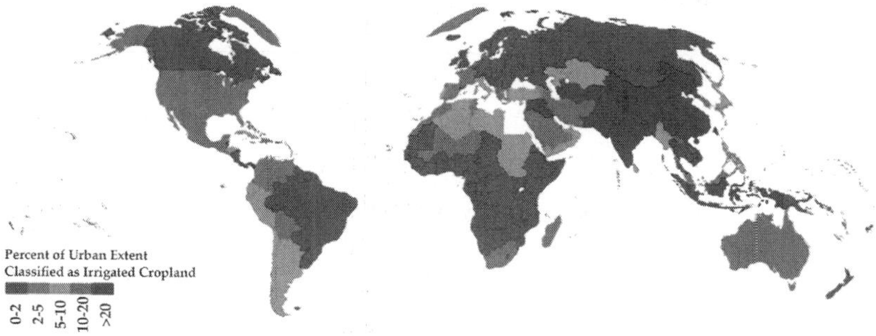

Figure 5.12: Illustrating where on urban farm irrigation is greatest by country (*Source:* https://www.npr.org/sections/thesalt/2014/11/12/363580928/40-percent-of-the-worlds-cropland-is-in-or-near-cities).

The primary constraint to irrigated urban agriculture is water. Availability of inputs, loss through crop damage and theft, including lack of labor are additional constraints. Even though lack of land, credit, and markets are thought to be the major problems, they are ranked lowest relative to lack of water (World Bank, 1999). This is instructive to the policy because most urban farmers do not use any credit and they do not view access to land as a constraint even though most of them are usually squatters.

5.8. WASTEWATER REUSE IN URBAN AGRICULTURE

Urban waste is considered as one of the most serious and pressing urban environmental problems. However, it appears that the main objective of most cities is to get rid of the waste giving little attention to the possibility of using it as an asset.

Exceptions exist however, because it is estimated that 10% of the world's wastewater is currently used for irrigation. Studies have shown that some cities such Santiago in Chile and Mexico City reuse all of their wastewater for irrigation. In South Africa, about 15–20% of the wastewater is reused in agriculture (Figure 5.13).

Figure 5.13: Illustrating a farmer in Mezquital Valley, Mexico using treated wastewater in his maize farm (*Source:* https://sanitationupdates.wordpress.com/2010/08/02/mexico-farmers-fear-loss-of-free-fertilizer-if-wastewater-treatment-plant-is-built/).

Urban agriculture occupies a special niche in urban waste management because it has substantive potential of reusing urban solid and liquid waste thereby contributing to overcome the waste problem while saving resources (Veenhuizen & Danso, 2007). Wastewater use in urban agriculture is a step towards improved water management for sustainable agricultural production because competition for water with urban household and industrial use is reduced. For example, Tanzania's National Urban Water Agency is reported to have expressed strong opposition to urban agriculture because farmers illegally leak fresh drinking water estimated at about 35% of the total supply. It is worth noting that about 80% of the land area in Africa is arid or semi-arid. Therefore, water is a constraint to sustainable livelihoods from agricultural production in these areas because it cannot be accessed in a cost-effective manner. In contrast, wastewater is available even during periods of droughts; its use would allow food production and the accompanying benefits, such as food security to continue throughout the year (World Bank, 1999).

However, if wastewater is not properly used it can be a threat to public health by contaminating soil and groundwater. In developing countries, the biggest threat to public health from reuse of wastewater are diseases caused by helminths (roundworms, hookworms and guinea worm), followed by microbial pathogens, particularly when untreated wastewater is used to irrigate vegetables or salad crops eaten raw (Mustafa, 1999). Wastewater has also some toxic components, particularly sodium, chloride and boron. The main negative effects of wastewater use on soil are increased salinity and reduced permeability; pores are clogged by grease in the water. Contamination of ground water in shallow aquifers is likely from microbial pathogens and nitrates in wastewater. Thus, urban administrators are bound to incur cost of disposing-off wastewater safely whether or not it is used in urban agriculture. From this, perspective urban agriculture can be a tool to achieve objectives of wastewater management and those of food security, jobs, income and sustainable urban environment (Mustafa, 1999).

The problems of wastewater reuse in urban agriculture are can be overcome because WHO in 1989 produced health guidelines suggesting ways of overcoming them. They guide policymakers in determining the appropriate wastewater treatment processes and in selecting suitable crops and irrigation methods (Blumenthal, 2001). For instance, the practice of restricted irrigation targets crops such as cereals, industrial crops, fodder, pasture or trees and excludes; crops likely to be eaten raw, sports fields and public packs. Irrigation should cease two weeks before date of picking and fruits on the ground should not be picked to reduce human exposure to contamination. In a similar manner, crops, soil, surface and groundwater areas most sensitive to contamination by wastewater need to be determined and excluded from this specific source of irrigation water.

Faraqui (2001) stresses that "even the most stringent treatment levels in the WHO guidelines can be met by a series of wastewater-stabilization ponds." These considerations would mesh well with the exercise of zoning urban land for specific types of urban agriculture as explained under land accessibility.

The guidelines single out sprinkler irrigation as inappropriate irrigation method with wastewater. It may even be too capital intensive for farmers. It is not clear if farmers from low-income areas are aware of the health risk of using contaminated water for irrigation. If they are aware, it is not clear if they have the incentive and the means to use safer alternatives. This is part of the nature and problems of urban agriculture practice as perceived by farmers targeted. Determination of awareness and incentive precede evaluation of the options available for overcoming the problem in a manner affordable by the farmers. It is notable that majority of women found cultivating previous dumpsite with contaminated soils in Kampala stated that they were aware of the health risk. Thus, they did not consume the produce in their own homes but sold it and bought alternative foodstuffs for their households (Ali, 2011).

5.9. SOLID WASTE MANAGEMENT AND URBAN AGRICULTURE

Solid waste is also another of the serious and urgent environmental problems facing cities in developing countries. The ubiquitous garbage heaps in developing countries, for instance, underscores the problem and the inability of the municipal authority to cope for various reasons. Rapid urban population growth rate has overwhelmed existing infrastructure, which is not expanded or kept in repair because revenue collection is hampered by the informal unplanned congested settlements, and the little collected is used for more urgent expenditure, such as salaries for a bloated labor force (Aosa, 1992). Municipal solid waste (MSW) of large cities in developing countries typically consists of 35- 70% organic waste. Organic waste represents a major challenge to MSW management for various reasons. They decompose, their nature and quantity has seasonal variation, they are usually mixed with non-biodegradable wastes, and there are practical difficulties in marketing compost products. Household kitchens generate the largest proportion of raw and processed waste. Other sources of MSW are gardening, urban agriculture, park and road maintenance, livestock keeping, and food processing. Bulk generators of raw wastes are green markets, parks, stables, and slaughterhouses and food processing industries, while large hotels and institutional kitchens are bulk producers of processed waste (Koont, 2017) (Figure 5.14).

Figure 5.14: An urban farming system using solid waste from the nearby households (*Source:* https://en.wikipedia.org/wiki/Urban_agriculture).

Composting and reuse techniques, including use for animal feed, fuel and in construction have been documented in Africa and Asia going back for a hundred years. Currently use of decomposed biomass in food and plant production in agriculture, horticulture, forestry and aquaculture is the means by which the greatest amount of urban organic wastes is reused. Therefore, urban agriculture has considerable potential in reusing MSW and in the process contributes to waste management and saving on resources (Pearson, 2011).

Peri-urban dairy and pig farmers around Nairobi and its environs for instance, constitute a high demand for raw and processed waste. This is an example of urban agriculture as a business contributing to urban waste management.

In low-income areas, livestock feed comes from farming activities and the local hotels. Since livestock farmers buy animal feed from farmers, sell milk to local families and finally sell manure to crop farmers, the nutrient cycle is complete (Richards and Godfrey, 2003). Therefore, this is one best example of waste and nutrient recycling in an urban setting. However, in this case, health concerns from keeping livestock in close proximity to people have also been harmonized with the waste management benefits. The observation by Richards & Godfrey (2003) indirectly suggests crop productivity and hence urban agriculture returns to farmers keeping livestock are different from that of farmers not keeping livestock.

Composting is the preferred method of processing organic wastes. However, research has shown that large-scale composting of urban organic solid waste (UOSW) is constrained by various factors. For example:

- Composting is expensive and therefore unattractive option relative to the low-cost open dumping currently used to dispose-off most UOSW in cities of developing.
- Small-scale compost plants run by nongovernmental organizations and community-based organizations in some neighborhoods, sometimes with the assistance of municipal councils in form of land, are financially unfeasible.
- Lack technical knowledge and the price for the small amount of compost produced is too high.
- Most households and institutions in developing countries lack the incentive to do the separation and urban waste managers have not found incentives to motive them.

- It is also not clear if composting of organic wastes reduces the health risks substantially as asserted because decomposition needs high temperatures (60°C), and several weeks of maturation to kill pathogens.
- Most cities, especially in Africa have not developed the institutional context for management of urban solid waste (Figure 5.15).

Figure 5.15: Famers in India urban area taking advantage of vermicompost in their farming (*Source:* http://www.vermico.com/india-farmers-use-vermicompost/).

5.10. FACTORS THAT DETERMINE LAND USE FOR URBAN AGRICULTURE

The nature of land use patterns attracted many scholars. Many studies have been carried out to find out what cause urban land uses to be located where they are; and whether there is a pattern of locational relationship between various land uses. Economic forces are just one of the many factors that combine to cause urban land use for urban agriculture. Various factors that affect land use will be discussed in detail in the following section of the chapter. They include economic factors, social factors, Government intervention, physical environment accessibility and ecological factors.

5.10.1. Economic Factors

Most researchers have attributed urban land use patterns to economic forces. Studies in American and western cities revealed that the land use relationship in urban areas tended to be similar in all cities. Cities tended to reveal structures that were somewhat identical in that they were divided into zones of similar economic activities. The relationship of economic activities and its spatial distribution usually took the form of easily distinguished areas having common functions depending on their distance from the city center.

Activities bid competitively in the land market for use of different locations. For instance, those activities that derive greatest benefits are able to outbid those which derive less benefit. Whereas the argument on competition is true, that of a resulting orderly arrangement may not be true, especially in smaller urban centers (OECD, 2003).

Goodal (2000) viewed the problem in the same way. He argued that due to competition for site by the various activities, in the long run an activity tended to locate where it enjoyed the greatest relative advantage in which case land was used for the highest and best use. He, however, rightly noted that this method of allocation did not consider the land use in the neighborhood. As a result, conflicts arose in the form of nuisance or encroachment which generated externality and defeated the very objective of land use.

According to Fox (2011), all cities cannot have the same kind of arrangement of activities due to other different forces that surround that particular city or town. It has been noted that in Africa, urban land uses are theoretically arranged in some form around the center of each city and that the cause of such an arrangement is mainly economic forces that tend to allocate each parcel of land to the highest and best use over time. The spatial arrangement of cities within each central place depend largely on economic forces and on the aggregate social tastes and preferences of each urban community as reflected through the local government actions and legislations in each urban government. He further observed that a systematic relationship exist between the location of retail shops, consumer service units and spatial distribution of residences; a tendency for the structure of land values themselves largely a result of competitive forces to mold the pattern of urban land use and influence spatial distribution of land uses within the city considerably.

Researchers argue that economic reasons in the periphery, coupled with speculation, causes land values in the peripheral areas to increase. In such cases, there is encroachment on agricultural land.

5.10.2. Social Factors

5.10.2.1. Social Values

On social values Firey (2001) did a study, in Boston involving an empirical investigation and development of a theoretical framework to identify the role of values in the evolution of land use patterns. He studied sections of Boston to determine how social values and ideals functioned with respect to past and existing land use patterns. In his study, he observed that the rationalists readily acknowledged the reality and effectiveness and social values, in spatial adaptation, but they made no attempt to incorporate the empirical concession into their theoretical system.

All factual departures from the kind of spatial order called for by the theory were lumped together into a loose category of limiting or complicating factors.

This category embraced customs, moral attitudes, taboos, political and administrative measures, cultural biases, traditional patterns and the like. All these were supposed to limit or complicate the natural competitive process but were not regarded as ultimate causative factors. Firey (2001) after studying locational trends in Boston's Beacon Hill, concluded that space may not only be a productive agent but also a symbol, and that people and groups chose locations not only in relation to market considerations but also in response to social values. He identified three kinds of influences which values exerted on land use. These he termed the retentive, the recuperative and the resistive. On the basis of his tests, he concluded that values were indeed self-sufficient ecological forces and that they had a very real causative influence upon land use. He further found out that residence to the northern end of Boston was a symbol of social solidarity and that of identification with Italian groups and Italian values. Those persons who identified themselves with Italian patterns tended to remain in the north end in spite of the deteriorated congested conditions which prevailed there.

Apparently, the effect which prevailed upon one's kind outweighed awareness of the slums undesirability as a place in which to live. Social values had thus an influence on land use which was not at all limited to areas with congenital physical and architectural characteristic.

In towns of Africa, social considerations may be strongly manifest in land use patterns. The major problem in articulating this is the fact that land use patterns in many areas are determined on administrative political considerations.

5.10.2.2. Religion and Culture

Religion and culture are also alleged to influence the shaping of urban land use patterns. Kingoriah (1999) observed that Mombasa's zigzag and irregular streets were not unlike those of many ancient cities where Islamic culture was predominant. This pattern was also prevalent in other ancient East African cities that had been heavily impacted by Islamic culture. He also noted that Kampala in Uganda developed as a result of opposing religious factors that were introduced during early colonial penetration times.

5.10.2.3. Population

Ratcliff (2001) whose major theoretical orientation in urban land use was mainly based on the interplay of market forces also recognized that economic activities served its population. These activities he noted were profoundly influenced by the nature of that population and that population trends therefore influenced future land use in urban areas in various ways and degrees.

5.10.2.4. Social Tastes and Preferences

Kingoriah (1999) in his study of "Policy Impacts on Nairobi's Land Use Patterns" found that within Nairobi, and very many urban areas, social tastes and preferences had played a great part in shaping the city structure.

These formed the component part of the city's political opinion and in most cases were the chief determinants of the structure of the city. He further noted that social preferences restricted land from changing use to the highest and best use. He concluded that economic factors were secondary to legal and social factors in the shaping of the structure of Nairobi.

5.10.3.Ecological Processes

According to Erickson (2002), the three ecological processes that offer a means of understanding of the patterning of towns are:

- Dominance.
- Gradient.
- Segregation.

5.10.2.1. Dominance

This is used in the sense of one area in the city bearing a controlling social or economic position in relation to other areas.

5.10.2.2. Gradient

This is a term that indicates the receding degrees of dominance from some selected dominant center to the more distant locations relative to that center.

5.10.2.3. Segregation

This is a process related to clustering. According to Erickson, Frederick, & Gutierrez (2002), the process of segregation and specialization involves the tendency of like units to concentrate on specific areas. Segregation refers to the sorting out of population into distinctive areas and specialization is a similar process but refers to the sorting out of uses, functions, activities rather than population. It has been argued that the degree of segregation or specialization achieved in an urban area depend on how long the process has been at work in the area and the degree of controls that are imposed to achieve these ends.

They further identified other ecological processes such as concentration and dispersion which constitute the initial process of massing and spreading population in urban areas in a region. This process explains the reason of the various urban centers and is to be distinguished from those processes which actually determine the internal organization of the centers. Another process is that of centralization and decentralization which refer to the congregation of people and urban functions in particular points and the reciprocal movement of people and activities to fringe areas or new satellite centers. A common result concentration process is the tendency for higher densities of population and intensities of use to occur in their inner parts of an urban area and the progressive decline towards the periphery.

Invasion and succession processes also identified by Erickson, Frederick, & Gutierrez (2002), are both associated with the foregoing sets of processes. Invasion refers to the penetration or encroachment of one population group or use type finally displaced the former. The main consequence of invasion is a break-up of the existing population and the land use mark up of an area. Succession is the culmination of the break-up process with the new achieving a complete displacement of the old. Thus urban growth and change involves not only the outward expansion of settlements but also internal reorganization to acknowledge and accommodate the constant changes which occur in the socio-economic circumstances of settlements.

According to Verma (1993), economic factors are constantly interacting with the social determinants which he refers to as socially rooted determinants of land use. He classified these two factors as the ecological and the organizational. He, however, described the ecological process as the process that describes the evolution and development of urban communities in space through the process of aggregation. Whereas these ecological processes may be experienced in large cities, this may not be the case in smaller towns.

5.10.3. Government Decisions and Policy

Some geographers suggested that cities have land use patterns that tend to resemble a series of roughly concentric zones based on land values, with the highest land values for the central business district. By and large most towns show this without government intervention but the pattern is usually lost after government. In guiding and controlling economic activity the government alters the relative profitability of different land uses as well as determining the economic framework which form the background to profit making decisions (Lichfield, 2001).

All government activities affect the patterns of land use in varying degrees. The laws relating to land use directly such as the planning acts are likely to have more effects on land use than say laws referring to the use of labor such as shops acts. Taxation and government tax expenditure is likely to have effects on land use. For instance, if the government tax a particular type of property and spent money on acquiring land for public use, these are more likely to have a greater effect on land use patterns than say the purchase tax on goods and the giving of pensions. According to Lichfield (2001), governments are forced to interfere on the urban land market because a land market left on its own devices could hardly be expected to produce an arrangement of land use patterns and development which would be socially acceptable.

5.10.4. Urban Land Use Planning

The granting of individuals access to the use of urban land is preceded by the allocation of such land between various uses. This allocation is done in accordance with a plan which thereby determines the urban development pattern.

Urban planning in most countries has not always been understood. In fact, some people at certain times have contended that urban planning is not necessary and that socio-economic factors automatically interplay to establish an optimum allocation of urban land between various demands. Another aspect of development control is through zoning. Through zoning governments try to maximize the social welfare by reducing the cutthroat competition between entrepreneurs and maximizing adverse external effects of land use into urban dwellers. The humble citizen is affected by this competition and therefore needs government protection.

According to Omosa (1998), it is wrong these days to think that land use in town centers is determined by the need to make as much profit by the landowners as possible. It has been argued that zoning plays a major role in determining the emergent land uses in urban centers. Zoning restricts landowners from using their land in any manner they please. This works as far as zoning regulations are strictly observed by both the landowners and the development control authorities. If inadequately observed, it could allow conflicts in land use. On the other hand, zoning practices works well where an urban center had abundant public land; otherwise it becomes difficult to enforce zoning and planning regulations in freehold land. It suffices here to note that planning in urban centers is racially motivated so that up to this day, racial districts – replaced by income groups are still conspicuous.

5.10.5. Physical Environment

Many types of physical handicaps exist in or around urbanized areas. These divert the direction of land acquisition and urban development in the direction where these handicaps are least or they do not exist. The expected pattern of city structure may then be modified accordingly. These and many distortions on the city structure result from the physical environment as it influences the land users' choice on where to locate their land using activity. Some of these environmental physical features include topographical features, climatic factors, the soils, and drainage system.

Climatic factors affect the development of land use patterns in an urban area. King'oriah (1999) asserted that although policy factors, especially the political interests of the European settlers may have been dominant in the determination of land use over the city, the climatic variation influenced the classes in power during the formative years of the city's structure; and made them accept a systematic method of residential land use allocation whereby expensive houses of the high income were found on the cool escarpment, while the hotter plains were occupied by the low income housing and industries.

He rightly concluded that in most East African cities the micro- and meso-climatic factors in the regions occupied by cities had the effect of influencing past decision makers with respect to where housing of various socio-economic groups would be located. In the highlands the locational dichotomy was between the breezy panaromic hillsides and the hotter valleys or adjacent plains. Climate therefore has a strong influence in causing the distortion of the land use pattern from the expected dictates of pure models with all their neo-classical simplifying assumptions. This, however, may not be manifested in smaller towns whose area of jurisdiction may be quite small as to have varying climatic conditions.

5.10.6. Accessibility

Accessibility is of similar importance for prospective purchasers may demand accessibility characteristics from their chosen sites. It has been seen that specialization reaches its peak in urban activity and that it depends upon exchange. Exchange necessarily involves contact between factors of production and producers, 'producers and wholesalers, wholesalers and retailers and consumers and so on. Accessibility sums the ease with which these contacts are made. A business has also to weigh the advantage of easy access to factor, inputs.

On accessibility in USA, Boer (2001) observed that increased accessibility was counterweighted by consequent changes in land use. People and institutions were operating on a larger scale because of and dependent on this new access through new infrastructure built to improve overloaded streets and roads.

5.10.7. Other Determinants

Sources of employment, infrastructural layout, and location of schools all play an important role in influencing the sum total of urban residential location decisions. According to King'oriah (1999), these act to distort the patterns of urban use and give greater conception insight into the economic variables. Existing development therefore acts as a factor determining use of urban land and the resulting land use pattern.

Land ownership is also an important factor determining urban land use pattern. In outlining the basic role of the then Physical Planning Department, it was stated in the 1970–74 Development Plan for Kenya that it was difficult to achieve the task of preventing wasteful encroachment of urban growth on agricultural land where land was held in small parcels especially in the municipalities. In such cases, it was argued, development was on a piecemeal basis and there were no development conditions as was the case with leased government land. The land being in the hands of many owners was developed at the owner's will and they could not be forced to develop it within a certain period for it would be impossible because of lack of finance. Thus, while one individual may put up a modern building complying with the building code, his neighbor may still continue to cultivate his land with only mud houses. This land use conflict was prevalent in the peri-urban areas.

One may, however, argue that since the Commissioner of Lands approves towns plans which control the use of all urban land, he in a sense controls the use of freehold urban land and moreover, the change of use of land from agriculture to urban on extension of an urban boundary must receive the approval of the Commissioner of Lands. The history of the evolution of any town should not be overlooked because it also contributes to the town's land use pattern (King'oriah, 1999).

6
CHAPTER

DAIRY CATTLE KEEPING AND AGRICULTURAL SUSTAINABILITY

CONTENTS

6.1. Introduction 150
6.2. Major Dairy Cattle Breeds 150
6.3. Sustainable Dairy Farming Practices 155
6.4. The Importance of the Dairy Industry 157
6.5. Milk Marketing 158
6.6. Factors Influencing Milk Production in Dairy Cattle 160
6.7. Incentives for Sustainable Dairy Farming 164

6.1. INTRODUCTION

Dairy farming is one field in the agricultural sector that has helped to revolutionize the lives of many people around the world. There is a lot that can be learned from this field, starting with the types of breeds to capitalize on, management practices, and the benefits that come with dairy farming practices. This chapter provides basic information that would be beneficial in one way or another. Some of the areas discussed include: major dairy cattle breeds; sustainable dairy farming practices; the importance of the dairy industry; milk marketing; factors influencing milk production in dairy cattle; and incentives for sustainable dairy farming.

6.2. MAJOR DAIRY CATTLE BREEDS

6.2.1. Holstein

This breed originated from Holland (which is in the Netherlands). It was first raised in the United States of America in the year 1852. Physically, this breed appears larger than other dairy breeds, and it has some black and white spots on its body.

Figure 6.1: Holstein breed (*Source:* https://cattleinternationalseries.weebly.com/holstein.html).

At birth, a health calf of Holstein dairy breed normally weighs roughly 90 pounds. On the other hand, under proper management, a fully-grown-up dairy cow would weigh about 1500 pounds.

Among all dairy breeds, this breed is the best in the total milk production. Per day one cow can produce about 60 pounds or roughly 8 gallons of milk. The butterfat content in the milk of this breed is normally 3.5% while the content of protein is normally about 3.0%. Studies have shown that this is the most reared dairy breed in the United States of America (for every 10 dairy farmers in the United States of America, 9 of them normally have Holstein breed in their herds).

6.2.2. Jersey

This dairy cattle breed originated from Jersey. It was first reared in the United States of America in 1850 s. Among all the dairy cattle reared for milk production, this is actually the smallest breed. Roughly a calf of this breed at birth would weigh 60 pounds, while an adult Jersey cow would weigh about 900 pounds. Physically, this breed is normally brown in color, and sometimes its color range from beige to almost black. Some breeds may also have some white spots or markings on their body.

Figure 6.2: Jersey dairy cow (*Source:* https://www.indiamart.com/proddetail/jersey-heifer-cow-16603147091.html).

In terms of milk production, this is an average milk producer. Per day it can produce about 45 pounds or 5.5 gallons of milk. The milk normally has a butterfat content of about 4.5% and protein content of 3.6%. Therefore, most people prefer its milk for butter and cheese making.

6.2.3. Brown Swiss

This breed is thought to have originated from Switzerland, and it is the oldest breed in dairy sector. It was first reared in the United States of America in the year 1869. It is actually one of the largest dairy breeds found in America. Its color may range from gray to brown. Most of the calves of this breed are normally born white in color, but as they age, they turn to brown or gray. Brown Swiss dairy breeds (Figure 6.3) are known for the following characteristics:

- Calm temperament.
- Long lives.
- Strong legs and feet.

Figure 6.3: Brown Swiss dairy cow (*Source:* https://www.brownswissusa.com/Breed/History/tabid/89/Default.aspx).

However, they have the longest gestation period (about 9 month and 15 days). This is actually 10 days longer compared to the other types of dairy cattle. Their milk production is on average, amounting to about 49 pounds (6 gallons) of milk. The butterfat content of Brown Swiss's milk is normally 4.0% while the protein content is 3.4%.

6.2.4. Guernsey

This is another very important breed in dairy farming sector. It originated from Guernsey. In the United States of America, this breed was reared for the first time in 1840. Body-wise, they are medium sized weighing about 1200 pounds. Its body is normally covered with white and brown spots (Figure 6.4).

Figure 6.4: Guernsey dairy cow (*Source:* http://www.guernsey.co.nz/why-guernsey/why-guernsey/).

Per day this breed is known to produce about 41 pounds (5 gallons) of milk. Its milk has a very high butterfat content of about 4.5% and protein of 3.4%.

6.2.5. Ayrshire

This breed originated from Scotland and was first introduced into American soil around 1832. Body-wise, this breed is medium in size weighing about 1200 pounds. The breed has red and white spots on its body skin. Most of these spots are normally small and jagged (Figure 6.5).

Figure 6.5: Ayrshire dairy breed (*Source:* http://www.ayrshire.co.za/p25/products/woolworths-ayrshire-farms-and-products.html).

This breed is known to produce about 43 pounds (5 gallons) of milk. Its butterfat content is about 3.8% while that of protein is about 3.2%. Some of its best characteristics include been fantastic forager and hardiness. In American, most breeds of Ayrshire are normally found in those coldest climates of Midwest and northeast states.

6.2.6. Milking Shorthorn

This is actually one of the oldest dairy cattle breeds in the whole world. This breed originated from northeastern England, and is believed to have been introduced in America in the year 1783. During those old days, this breed was used for meat and milk production and it also offered transportation for the American pioneers.

On average, an adult Milking Shorthorn cow would weigh about 1300 pounds. Physically, the body of this breed may be covered with red-white, red, or roan color (Figure 6.6).

Figure 6.6: Milking Shorthorn dairy breed (*Source:* https://www.petmapz.com/breed/milking-shorthorn-cattle/).

They are known for their docile temperament and can produce up to 39 pounds (4.5 gallons) of milk in one day. The butterfat content of its milk is normally about 3.6% while protein is about 3.1%.

6.3. SUSTAINABLE DAIRY FARMING PRACTICES

6.3.1. Housing

Most dairy farmers currently use what is referred to as free stall rearing system. This is a type of housing system which allows the dairy cows to consume their feeds and sleep whenever they feel like (Figure 6.7).

Figure 6.7: Illustrating dairy cows been reared under free stall system (*Source:* https://www.dairynz.co.nz/farm/off-paddock-facilities/freestall-barn/).

This type of housing system also allows the dairy cows to have clean water and fresh feeds at all times. For the comfort of the dairy cows, some farmers normally install some rubber mattresses for them to lie on while resting. However, farmers are always advised to make sure that the barns are well ventilated in order to ensure there is free flow of fresh aim in and out of the barns.

6.3.2. Feeding

6.3.2.1. Maize Stalks

In most parts of the world, it is a common agricultural practice, especially among the small-scale farmers to remove the green maize ears for home consumption. The remaining green maize stalks are normally used as cattle fodder.

Maize as a crop is extensively grown in most parts of the world. It is rational therefore to expect a fairly large quantity of maize stalks available for livestock feeding on these small-scale farms. However, there has been little research work on the utilization of maize stalks by livestock.

Abington (1992) has shown that the digestibility of maize stalks declines with age and varies according to many other factors, such as environmental temperatures and cultural practices. According to him, young maize stalks are highly digestible and low in percent Cell Wall Constituents (CWC) whereas mature maize stalks are less digestible and richer in percent CWC (Figure 6.8).

Figure 6.8: Farmers grinding maize stalks for animals (*Source:* http://www.dhahabu.co.ke/2016/08/30/bunkers-feed-preservation-earn-muli-sh12-million/).

6.3.2.2. Maize Cobs

These are the by-products of maize ears after the removal of mature maize grains. Maize cobs can be a very useful feedstuff, mainly as a source of roughage to the ruminants. Globally, there is a very high volume of maize cobs which go to waste, and could be used for livestock feeding (Gebremedhin, 2002). However, studies have shown that some small-scale farmers and a few large-scale farmers grind maize cobs together with maize grains for livestock feeding.

6.3.2.3. Sugarcane and Sugarcane Rejects

Sugarcane thrives successfully in the warm humid high potential areas. The crop is grown on both small-scale farms and on plantation estates. Sugarcane has been used and is being used for livestock feeding in many parts of the world. The potential of sugarcane and sugarcane rejects as a contributor to livestock development in the tropics is strongly associated with its high energy yields per unit area of land (Lionel, 1975). The total digestible nutrients (TDN) per unit area of land of sugarcane surpass the yield potential of cereal grains and tubers.

Sugarcane is normally at its optimal nutritional value in the dry season; hence its use eliminates animal weight losses which coincide with dry season. In view of these spectacular characteristics, sugarcane considerably increases animal off-takes per unit of time. Lionel (1975) thus found out that slaughter weight of beef steers was reached at 2.0 years of age, compared with to 3.0 years of age for steers on pasture.

Sugarcane tops constitute an important fodder to cattle although at present they are not fully used, in spite of the scarcity of green fodder in most countries. It is lamentable that in spite of the fairly huge tonnage of cane tops available sugarcane growing countries, no work has been done to assess their nutritive value and the practicability of feeding them to the ruminant animals.

6.4. THE IMPORTANCE OF THE DAIRY INDUSTRY

Milk is important in a country's economy both as a source of animal protein and in terms of its gross marketed value. Milk, at the farm level, is used for home consumption and calf rearing, and the surplus is sold either in the rural markets or to processing plants. The dairy sector provides a source of income and employment for the dairy farmers and their employees, as well as employees of the processing and distribution sectors. The exact number of people employed in the dairy sector is, however difficult to determine. This is because most of the marketed milk production in a country, especially in developing regions, is from the smallholder sector where dairying is but one of the many enterprises undertake in mixed farming operations (Abington, 1992).

Milk products are among the commonly consumed items in the diet of most countries' population and are particularly vital as far as the nutrition of young children is concerned. Milk dominates the pastoralists' diet and, from a nutritional point of view, is or of the major sources of animal protein.

Apart from being a basic food, milk is a raw material in the food processing industry, for example, in the processing of baby foods and chocolates. Dairying also supplements the market by producing cattle and calves for the market. Dairy farming is also important as a source of manure which may be applied to crop farms as a substitute to the more expensive inorganic fertilizers.

Dairy farming also has implications for foreign currency reserves. Local production and processing play a role as a source of foreign exchange savings through substitution of imported dairy products.

The overall importance of dairying is therefore, much more than can be deduced from a mere consideration of local sales milk and milk products (Gibon & Mihina, 2003).

6.5. MILK MARKETING

Globally, milk is normally marketed through two distinct channels, which are:

- The regulated "formal" channel.
- The unregulated "informal" channel.

6.5.1. Formal Channels

These include the cooperatives and other licensed dairy processing plants. They are formal in the sense that they are under government regulation.

Formal channels normally play an important role in the marketing of the smallholders' milk. Milk is delivered by the farmers to specific collection centers where it is weighed before being collected by hired or Cooperative Society vehicles. According to Muriithi (1990), most of the milk collected by societies from the farmers is first sold locally to individuals and institutions such as schools, hospitals and hotels along the market route. The remaining amount is then sold to processing plants for further processing.

6.5.2. The Informal Channels

These marketing channels refer to those which operate without the government license. This comprises all milk sold other than to processors. This includes sales to neighbors, "hawked" milk and unprocessed milk sold by cooperatives in their locality.

According to OECD (2003), pricing policy for milk in a country is essential because it encourages dairy producers to maintain a satisfactory and constant supply, as well as to protect the consumers.

Prices in the informal market usually tend to be higher than those in the formal market. This could occur either as a result of demand outstripping supply or due to some institutional problem(s) such as high-cost of production. It is thus worthwhile to examine the industry in order to determine what the causes of these high prices are (OECD, 2003).

6.5.3. Role of Milk Marketing

Milk marketing helps in linking farmers with the rest of the economy as it enables them to sell off their surplus produce while obtaining the inputs required for their production process as well as the final goods that they do not produce but requires for their consumption. There are three important marketing channels that require to be developed in order to offer incentives to dairy farmers with intent of making them more productive and integrated in the whole economy. These marketing channels include:

- Channel for dairy farmer produces.
- Channel for obtaining consumer goods and services that the dairy farmers require but do not produce.
- Channel for acquiring farm inputs.

The existence of an efficient and flexible agricultural marketing system makes it easier to achieve a smooth transformation of the agricultural sector in line with national development strategies of any developing country. Mosher (1966) identifies what he calls the five essential requirements that constitute a helm for agricultural sustainability in dairy farming enterprise. These include:

- Markets for dairy farm products.
- New dairy farm technology.
- Local availability of dairy farm supplies and equipment.
- Adequate incentives for the dairy farmers.
- Transportation facilities for dairy produce.

For sustainable dairy farming practices, such a system should possess the following characteristics:

- It should have the potential and the capacity to cater for increased output;
- It should offer critical and useful information to the dairy farmers in order for them to offer the goods and services required by the consumers at prices they are willing to pay.

6.6. FACTORS INFLUENCING MILK PRODUCTION IN DAIRY CATTLE

Studies have shown that the level of milk production in dairy cows is as a result of many factors within the cow and environment. In most cases, dairy farmers can alter many factors in order to improve milk yield and improve its composition, while some factors are beyond the farmers' control. It has also been shown that most factors which influence milk production also contribute to the heterogeneity of variance of milk yield between herds. Some of these factors include:

- Age at calving.
- Period of calving.
- Lactation length.
- Dry period.
- Calving interval.
- Body weight at Calving.
- Genetic parameters.

6.6.1. Age At First Calving

Milk production in dairy cows is influenced by the age of the cow at first calving. It has been established that too early calving may slower body development of cows. For instance, Friesian dairy cows should not calve before they are two years old. However, research has shown that most Friesian cows are normally mated and calve at this age provided that they have attained two-thirds of their mature body weight and are in good condition (Kebreab, 2013). Therefore, farmers can used both the age and the weight of the dairy cow to determine the right time for mating.

The increase in milk production with parity depends on the selection made, and the actual age. Most of the time, selection of the dairy cattle tends to make older cows appear to have higher yields than the younger ones, if the "gross comparison method" is used.

There are several methods that can be used to correct milk yields for parity effects. According to research, multiplicative factors are preferred to additive factors. Regression techniques have also been used. Where many herds are involved, simplified herd level age correction factors can be used.

Parity and season of calving interactions may exist in some situations and have to be allowed for in data correction. It seems important to test for parity by season interaction when deriving parity correction factors. It is worth noting that the level of production of the cows in the herd may affect the age correction factors. Sutherland, Wilson, & Zagata (2014) showed that a highly significant relationship between age of cow and yield remained in the conventionally age corrected records at all the levels of management. The bias appeared to be less marked in the well-managed herds. Where different levels of management, breeds and herds are involved, it is safer to derive correction factor on a within environment, breed and herd basis.

6.6.2. Period of Calving

Management, climatic and genetic factors are the major causes of yearly variation in dairy cattle performance. In the tropics, rainfall is the major climatic factor that influences pasture quality and availability. Animals that calve down during unfavorable years with reduced pasture supply perform poorly (Gibon & Mihina, 2003). However, supplementation and better management improves herd performance. Due to differences in supplementation and management levels between herds, variations in herd performance still occur. However, the effects of management and climate are usually confounded.

Subsequently, it is significant to recognize and categorize the real causes of year effects. Reports on the influence of season on production performance of cattle, especially those raised in the tropics are inconsistent. Some studies have reported significant effects of calving season on milk yield while others have found non-significant season effects.

Supplementation of animals during the dry season may lead to non-significance of the season effects. It has been observed in Kenya that although season of calving did not have a significant influence on milk yield, cows calving in the short rainy season had the highest milk yield while those that calved in the long rainy season had the lowest (FAO, 1993). This is attributed to the small part of lactation supported by the high-quality pasture for 'cows which calved in the long rainy season. Those that calved during the short rains are normally advantaged by the long rains which coincided with their lactation period.

6.6.3. Lactation Length

According to the observations made by Gebremedhin (2002), there is an always high correlation between lactation length and milk yield. The degree to which lactation length is associated to milk production normally depends not only on the breed but also on management.

Many research workers do not correct milk yield for variation in lactation length despite the fact that lactation length is significantly correlated to milk yield. They argue that lactation length is genetically determined and that correcting yield tends to reduce genetic variance. It may be justified to correct milk yield for variation in lactation length if the heritability of lactation length was almost zero (Abington, 1992).

6.6.4. Dry Period

It has been observed that most dairy cows with the shortest dry periods normally have the highest milk yields. Most tropical cattle tend to have long dry periods and they therefore get more than adequate rest periods.

The influence of dry period on the total milk production has not been shown for many tropical breeds. It would in general, appear that there is very little accuracy to be gained by corrections for dry period of dairy cows, especially those found in tropics.

6.6.5. Calving Interval

Calving intervals appear to influence milk yields only when they are short. When they are very long, their influence is reduced as each cow gets more than a sufficient rest. Calving interval has a low heritability. Therefore, correction factors should be used before cows are selected. Dry period and lactation length which are components of calving interval have to be considered when correction factors are being devised.

6.6.6. Body Weight at Calving

It has been argued that a large dairy cow is not desirable because she has high maintenance requirements. Others researchers argue that comparison of efficiency of milk production, based on output per unit of input is more significant than a comparison of sizes of cows, since, in general, larger cows tend to produce higher yields than smaller ones.

Studies have shown that for every 50 kg increase in weight at calving, milk yield is likely to increase by 60–80 kg per annum. However, it important to note that above relationship depends on breed, herd, age and the method of analysis (Kebreab, 2013). It can be argued that heavier cows are able to produce more milk because of their ability to ingest and metabolize larger amounts of feed and despite their higher maintenance requirements; the extra energy is turned into milk.

6.6.7. Genetic Parameters

Both phenotypic and genetic differences between different dairy cattle breeds are considered by geneticists as the raw material for improvement. The impact of heterogeneity of variance on animal evaluations and selection has been examined by several researchers. Everett (1982) found that among herds of equal genetic merit, herds with greater variance had a higher proportion of cows achieving elite status than herds with lower variance. Powell et al. (1983) reported similar findings. Also noticeable that among herds of equal sire merit, herds with higher production level had greater variance attributed to better nutrition and management.

In a simulation study on selection among herds with heterogeneous variance, Hill (1984) reported the fraction of animals that would be selected from the more variable groups under differing intensities of selection. The proportion selected from the high variance group increased as both selection intensity and standard deviation increased.

Vinson (1987) noted that heterogeneous herd variance causes genetic evaluations for high producing cows in high variance herds to be exaggerated.

Using evaluations based on progeny tests, Wilhelm & Mao (1989) showed that young bulls selected from herds with low milk yield variance were genetically superior to those selected from herds with high variance. They also observed that predicted transmitting abilities for young bulls selected from low variance herds were less biased, while those of bulls selected from high variance herds were inflated probably due to genotype–environment correlation.

In cow evaluation, information about an individual cow and its relatives is combined into a cow index and is expressed as breeding value (Brotherstone and Hill, 1987). Different methods are used worldwide in cow indexing. The use of a cow genetic index for dam selection is complicated by heterogeneous variances arising from differences in levels of production between herds (Wilhelm and Mao, 1989). Yet for its efficient use, the index has to reflect the genetic merit of the cow correctly irrespective of the level of herd production.

Heterogeneous phenotypic variation is mainly attributed to genetic and environmental causes (Short et al., 1990). Differences between herds in culling levels and breeding skills make it difficult to eliminate genetic inter-herd variation. However, the use of artificial insemination sires across herds or assuming a fixed amount of genetic variation in herds may eliminate inter-herd genetic differences.

Environmental differences may be eliminated by within herd corrections (Mosi, 1984). These findings suggest the need to account for heterogeneity of variance to minimize the effects of variation in herd yield level on the index values. This would overcome the bias in evaluations and remove disproportionate selection that results from heterogeneous variances.

6.7. INCENTIVES FOR SUSTAINABLE DAIRY FARMING

To improve production in dairy sector, there is need for a total transformation of the agricultural practices. Transforming traditional dairy farming into a highly productive dairy sector depends on the investment made on agriculture and form it takes make it profitable (Gebremedhin, 2002). In most countries around the world, dairy farming is treated as a source of economic growth which can act as an engine of development, but the form of investment is important for the realization of this goal. Studies have shown that incentives to guide and remunerate dairy farmers are seen as significant elements of investment to increase dairy farming productivity and sustainability.

There is little adoption to change and most dairy farmers are more secure about what they know about traditional agriculture than embracing and learning how to use new factors of dairy production. The types of risk and uncertainty about yield associated with the advance in knowledge are of real concern to most dairy farmers who produce so little barely enough to meet subsistence requirements.

It has also been note that most dairy farmers are always resistant to change because introducing new factors mean dealing with problems of risk and uncertainty associated with dairy farming productivity of the new factors. The rate at which old-style dairy farmers assume to new factors in dairy farming is therefore subject to the allowance for risk and uncertainty.

According to Gibon & Mihina (2003), revolutions in dairy farming are presently restrained by lack of satisfactory incentives. The state of incentives is such that in many countries it is unprofitable for dairy farmers to start modernizing investment that would increase the productivity of dairy farming. An incentive in this case is the product of economic information from which the farmer derives his expectations. The cost expectation covers the rate of interest, rent, on land, payment for equipment, labor and the farmer's time devoted to dairy farming. On returns, expectations encompass the value placed on the dairy farm products to be used in the farm household and the expected price of the product to be sold. Optimum economic incentives offer the information that leads producers to allocate resources in ways that result in maximum production.

It has been noted that increased investment in mechanized equipment alone is not enough to increase milk production in dairy farming, rather, better management information and utilization of resources should be equally emphasized if the potential grains are to be realized. This according to the suggestions given by Abington (1992), means that no increase in production can be realized by reallocating the existing factors of production.

It can be concluded that most dairy farmers are normally challenged in terms of the total quantity of resources they are able to use by market imperfections. Most dairy farmers, especially in developing countries are inefficient in resource allocation. They are also risk averse and employ few resources in more risky dairy farming activities (Abington, 1992). Therefore, there are several steps that can be taken both at local and national levels in order to enhance their dairy farming productivity. These include:

- The government should take on some of the farmers' risks by guaranteeing minimum expected return.
- The government could also reduce the gap between buying and selling price of dairy products.
- Maximization of dairy farm output at local market prices, subject to resource constraints like labor and land.
- Restructuring the extension policy and formulating pricing policy to consider dairy farmers returns in dairy farming. This can be achieved by considering good quality dairy products to increase the demand for the final product.

7
CHAPTER

BEEF CATTLE FARMING AND AGRICULTURAL SUSTAINABILITY

CONTENTS

7.1. Introduction 168

7.2. Beef Meat Demand and Supply 168

7.3. Challenges in Beef Cattle Production 170

7.4. Markets for Beef Cattle 171

7.5. Factors Affecting Beef Farming 175

7.1. INTRODUCTION

This chapter provides an overview of some important issues in the beef cattle farming, both locally and internationally. Specifically, meat demand and supply aspects are discussed in this chapter. Further, challenges and factors affecting beef farming industry are also discussed in this chapter. Markets for beef meat are also outlined briefly under this chapter.

7.2. BEEF MEAT DEMAND AND SUPPLY

Most beef cattle products, including meat, are important sources of nutrients, such as proteins in human nutrition, and micronutrients that are vital for physical and cognitive development in children. According to the FAO (2009), global demand for food is expected to increase by up to 70% by 2050. So as to meet this demand, it is estimated that meat production should increase from some 229 million metric tons in 1999 to about 470 million metric tons in 2050.

According to FAO (2005), beef production globally, amount to about 40% of the livestock output. In 2009, the total beef output was estimated to be 62 million metric tons. The United States of America (USA) was the leading producer of beef, supplying 19%, which amounted to about 11.9 million metric tons of the total output. Brazil was second with 15% amounting to about 9.1 million metric tons, followed by China at 10% amounting to about 6.1 million metric tons, Argentina with 5% amounting to about 2.8 million metric tons and Australia with 4% amounting to about 2.1 million metric tons) in 2009.

On average, these five main producers supply about 53% of total beef output, while the EU produces a further 13%. However, the growth rate in beef output from the five countries fell from about 11% per annum during the period 2001–2005, to only 1% in 2005–2009. Beef output in the EU also declined during this period (FAO, 2005).

Studies have shown that total beef output is expected to drop in future. This may be attributed to the emerging competition for land and other inputs from bio-fuel generation. However, human population in the world is expected to increase from its current level of nearly 7 billion in 2011, to about 9.1 billion by 2050 (UN, 2009). More than 90% of the anticipated increase in population will likely occur in developing countries, including sub-Saharan Africa (SSA) where the annual population growth rate is anticipated to be about 1.2%. The expected rise in population, together with urbanization and possible changes in expenditure caused by growth in incomes, are expected to drive demand for livestock products upwards (FAO, 1993). It is estimated that the average annual per capita consumption of meat (including beef) in developing countries will increase from some 28 kg in 2002 to about 44 kg by 2050 (Thornton, 2010).

The need to meet expected increases in demand for meat is coupled with challenges such as competition for resources between enterprises, and concerns to reduce greenhouse gas emissions in livestock food chains. Greenhouse gasses such as methane and nitrous oxide are considered to be major causes of global warming (commonly referred to as climate change) that is associated with adverse effects on the environment, including water pollution and loss of biodiversity. It is estimated that livestock production systems contribute about 25% of greenhouse gasses globally (National Research Council, 2010).

According to FAO (2009), there is often a rather extreme argument in some environmental debates that one option for the world to manage global warming is to stop livestock production. Nonetheless, considering the significant role that livestock play in human nutrition and livelihood enhancement, there is need for balanced interventions.

Normally, livestock production supports the livelihood of over 65% of the rural population in most countries around the world. It contributes between 15–30% of their agricultural Gross Domestic Product (GDP), provides food, draft power, manure, serves as a form of capital investment and provides cash income in times of need, serves as means of transport for goods and services, and livestock are often used in various African socio-cultural ceremonies, e.g., bullfighting contests (Abington, 1992).

Additionally, in the Sub-Saharan African region, where more than 70% of land in pastoral areas is arid or semi-arid, and therefore largely unsuitable for crop farming, livestock production is often one of the most viable enterprises in such areas (AU-IBAR, 2010). However, the African, Caribbean and Pacific Group of countries produce only 4% of total meat output in the world, and they have relatively low productivity. For example, the average slaughter weight of cattle is less than 170 kg in Africa, while for most developed countries it is over 400 kg (Abington, 1992).

Therefore, it is important to change the ways in which livestock farming is done and embrace the new technologies that are currently employed in order to enhance productivity and sustainability of beef cattle and other livestock rearing.

7.3. CHALLENGES IN BEEF CATTLE PRODUCTION

According to FAO (2005) report, it has been noted that the production and productivity levels of beef cattle in the world are generally low. These low levels of beef production are attributed to the following elements:

- Slow rate of growth of the total beef cattle population.
- High calf mortality.
- Poor management of beef cattle.
- Slow rate of maturing.
- A small final live weight.
- A low meat yield.
- Periodic heavy losses from drought.

The heavy calf mortality is in a large part responsible for the very slow rate of expansion of the total beef cattle population in the world. Apart from reducing the size of a surplus which might possibly be marketed, the slow rate of growth limits the number of locally produced breeding animals available for development schemes.

The high mortality among calves is superimposed on a very low calving average, and these two factors combine to hold back very considerably the growth of cattle numbers in the world. Both production and productivity suffer as a result. Losses of calves account for 20–25% of the natural annual production of animals (Abington, 1992).

It has been proved that under improved conditions the level of calf survival may significantly be raised. The relatively small number of calves which survive the first few months of life are often weakened and later stunted by the rigors of disease and reared under often quite unsatisfactory systems of management. The pressures of the physical environment such as lack of water, shortage of grazing, and poor management practices such as inadequate feeding of calves, delay the normal maturing time by as much as 50–60% so that beef cattle may take up to five to six year to reach full growth (Gibon & Mihina, 2003).

The slow rate of maturing alone considerably reduces the turnover per unit area of pasture and brings down the level of productivity per acre. Bearing in mind the largely non-commercial nature of most stock-keeping in some countries, it is largely academic to speak of the level of production and productivity in terms of animals reaching any cash market, but the level of production under current methods reveals the scope for considerable improvement as this sector becomes commercialized.

At maturity, most beef cattle, especially those reared pastoralists weigh an estimated 560 lbs. This figure is by no means an overall average. As it is often common for only the worst animals to reach the market, the figure is probably an under-estimate of the true position. Despite this, the figure of 600–800 lbs. which might result after the adjustment for a generally better animal plus the weight lost in movement to sale is still considerably below the 1000 lbs. or more which characterizes beef animals in some countries. The low weight reflects the poor plane of nutrition, disease problems and mismanagement (Kebreab, 2013).

7.4. MARKETS FOR BEEF CATTLE

Beef cattle in most countries around the world are normally marketed through four main types of market. Each market type is distinct in terms of location, consumer hinterland, purchasing hinterland and, method and volume of trade. The four principal market types are:

- Rural Butchers.
- Local Government Markets.
- Resale Markets.
- Ranch Markets.

7.4.1. Rural Butcher

Most rural butchers normally operate in the countryside, usually at the local produce market where he may be a paying stallholder. Butchering is a part-time activity and prior to market day; the butcher leaves his work on his smallholding and visits farmers in the local area in search of cheap beef cattle for slaughter (Figure 7.1).

Figure 7.1: A typical rural butcher (*Source:* http://berniesafricanodyssey.blogspot.co.ke/2013/03/nyama-choma-time.html).

Studies have shown that the prices paid for these animals are generally lower than those prevailing in the organized markets and, partly for this reason; beef is often 25–50% cheaper in the countryside. The reason to why most local farmers are usually willing to part with animals at this lower price revolves principally around a fear of veterinary inspection at the county markets. Where cattle are in very short supply, the rural butcher will often make-do with small stock, for which there is sometimes a preference and a higher unit return. If peasant farmers in the area are willing to dispose of animals cheaply, the butcher will purchase over his immediate requirements and maintain the surplus cattle on his small farm until needed (Lichtfouse, 2014).

The existence of the rural butchers is dependent on the prevailing level of rural income. When money is readily available it is common for the demand for meat to rise, and so the number of butchers varies seasonally with both the volume of the crop harvest and the level of return from the crops. At times of hardship it is not uncommon for butchers temporarily to abandon their commercial interests and return to the production of subsistence crops until more money is circulating in the local area.

Where the village butcher operates close to an administrative center such as a sub-county headquarters, provision is made for the hygienic slaughter of animals on a local authority concrete slab where animals can be killed under the supervision of veterinary staff. This considerably increases the risk taken by the butcher, though at present, the general absence of veterinary inspection at the village level and the freedom from fear of carcasses condemnation encourages the small butchers to continue their trade with only a limited risk on their very little capital (Veenhuizen & Danso, 2007).

7.4.2. Local Government Markets

Most local government beef cattle markets are usually organized. They serve four major functions:

- Commercializing the livestock sector and spread the money economy
- Assuring a regular supply of meat to the growing townships
- Offering a more lucrative alternative to the rural traders
- Proving a source of revenue to the local government

The market network covers, with varying degrees of intensity, and it is usually under the control of the local district administrations. Local government markets are different from the rural butcher system in scale of operation and in not selling meat (Figure 7.2).

Figure 7.2: An example of local government market for beef meat (*Source:* https://psmag.com/social-justice/support-industrial-slaughterhouses-local-meat-76120).

The market formations offer a place where cattle may be traded for consumption elsewhere. The local government cattle trading centers have two main types of trade:

- Supplying the urban and high-income peri-urban market within the district.

- Supplying animals for transfer to urban or high-income peri-urban or rural areas in other parts of the country.

Urban butchers, traders, both buyers and sellers and peasant farmers come together to do business. The market may be well equipped with fencing and water and a hut for the resident veterinary officer who must preside over any sale.

7.4.3. The Resale Markets

It was observed by Abington (1992) that these markets sell directly at the railhead for immediate slaughters. In most places, the butchers from the high-income peri-urban area attend the daily market and buy animals which are then immediately slaughtered. The meat is transported by vehicle to various retail outlets within a radius of 30 miles. The need for transport and the very high price prevailing at the market may exclude the small-scale village butchers.

7.4.4. The Ranch Markets

Most ranch sales in different countries are conducted as auctions at regular, advertised intervals. Sales differ at the ranches in being organized on a basis of known weight and quality grade introducing the basis of quality/price differential which is either lacking or only crudely estimated at the other forms of market (Figure 7.3).

Figure 7.3. An example of ranch market (*Source:* https://www.weeklytimesnow.com.au/agribusiness/livestock-sales/prime-cattle-sales/prime-cattle-market-remains-firm-as-restocker-categories-lift/news-story/7f4524e8729a1d7f4bf44ef95ac0899b).

7.5. FACTORS AFFECTING BEEF FARMING

The significance of beef farming productivity cannot be overemphasized. Just like any other types of meat, beef is an exceptional source of protein, vitamins and minerals which are necessary for the growth and health of the human body. Beef is one of the most demanded meats in the world. Geometrical increase of population, particularly in developing countries, is one aspect that underlines the importance of increasing beef productivity programs so that beef output corresponds with the rising trend of population (Figure 7.4).

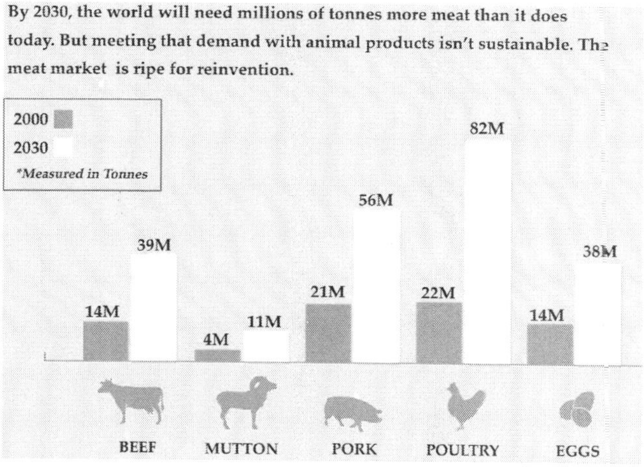

Figure 7.4: The demand for meat in the world (*Source:* https://timzimmermann.com/2013/04/05/meat-culture-americans-eating-less-meat-the-world-eating-more/meat-consumption-2030/).

Beef is also economically important both nationally and internationally. Beef is a lucrative export item in a country's foreign exchange earnings and today international trade handles more 15% of the world's meat production and one half of it is beef.

There are several factors that affect beef farming in most area, especially in tropics. For instance, studies have shown that the behavior of beef cattle on pasture depends on heredity, environmental stimulation and other factors (Gibon & Mihina, 2003). Among the important environmental factors are temperature, humidity, rainfall and sunlight/darkness. Other factors are bovine diseases, social attitudes, water resource, pasture equality and quantity, breed differences and general management practices such as kraaling, stocking rates and supplemental feeding.

7.5.1. Bovine Diseases

Most of the major diseases occupy a distinct geographical area either as a result of limits set by the physical environment on the host, vector and/or parasite, or as a result of the recent activities of man.

Ecological conditions favoring the spread of infections such as pleuro-pneumonia and trypanosomiasis exist in most countries, especially in Africa and other developing countries (Kebreab, 2013). As a consequence, strict control must be exercised to prevent the spread of the more virulent forms of disease. Possible expansions of infected areas through trekking for water and grazing in the dry season pose a considerable threat.

The diseases of cattle commonly lead to

- Abortion.
- Stunting.
- Delayed maturing.
- Carcase condemnation.
- A high level of mortality.

Although all these characteristics are common of stock in most parts of the world, many show a strong regional concentration.

Debilitation and high mortality exercise a considerable influence on the production of cattle and the productivity of both animal and unit area of pasture. Further, the diseases pose a considerable threat to development schemes based on exotic and crossbred animals and to the prospects for exporting meat to developed countries (Abington, 1992).

In Africa, bovine diseases can be classified into two main groups:

- Epizootic.
- Enzootic.

7.5.1.1. Rinderpest

This is a virulent virus disease which has in the past caused enormous loss in a short space of time. With the introduction of vaccines, the disease has been successfully isolated, though it is not yet possible to eradicate it altogether.

7.5.1.2. Pleuro-Pneumonia

This disease has undetected carriers much in the way that typhoid has carriers among human beings. This makes it difficult to confine the disease to one geographical area, especially as there is no form of prophylaxis. Once introduced, the disease will spread rapidly where animals are brought into close contact, and in order to control and contain the disease, immediate slaughtering is necessary for every beast which reacts positively to testing (Figure 7.5).

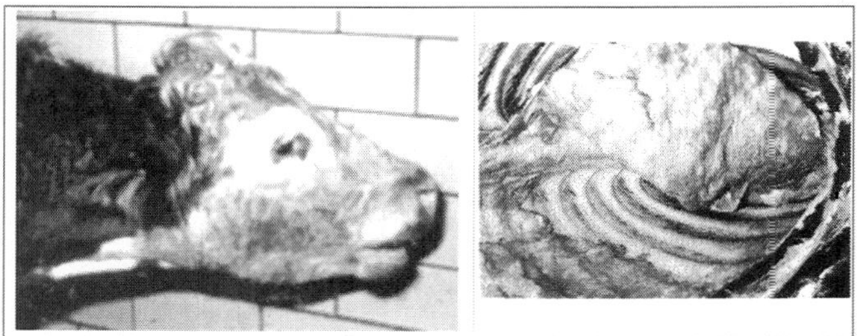

The extended neck and head is due to respiratory distress and coughing.

The thoracic wall is reflected; there is excessive fibrin the parietal and visceral pleurae and a cloudy fluid in the thoracic cavity.

Figure 7.5: A cow showing signs of Pleuro-pneumonia (*Source:* http://lrd.spc.int/ext/Disease_Manual_Final/a160__contagious_bovine_pleuropneumonia.html).

Pleuro-pneumonia is the principal cause of the rigid quarantine system and on-site slaughter requirements which so strongly influence the spatial and chronological pattern of beef cattle supply in most countries around the world. Although the disease causes very little actual mortality at present, it must be strictly contained.

7.5.1.3. Trypanosomiasis

This is the bovine form of human sleeping sickness and is transmitted by both tsetse and biting flies. The disease results from the introduction of the parasite or trypanosome during the blood meal of a tsetse or biting fly. The vector or fly feeds from favored hosts among both cattle and wild game and the latter, in cases such as bush pig, acts as an immune reservoir for the disease (Figure 7.6).

Figure 7.6. Anemic and weakness are some of the common signs of trypanosomiasis in beef cattle (*Source:* http://www.veterpedia.net/meat-inspection-spec/378-cattle/trypanosomiasis/judgment-differential-diagnosis/718-trypanosomiasis-judgment-differential-diagnosis.html).

If the level of infection in an area is moderate to heavy, then it is impossible to herd domestic cattle. This is because the mortality rate would be too high, but if light, the keeping of stock is possible under prophylaxis.

As small stocks are immune to trypanosomiasis they form the main element of meat production and supply where the fly is present in strength. The distribution of the savannah forms pallidipes and morsitans are principally responsible for disseminating the more virulent congolense or vivax forms of trypanosome, and were seen to be an important factor in limiting cattle production in most parts of Africa.

The distribution of tsetse and trypanosomiasis is strongly influenced by factors affecting the vector and the parasite. For instance, the vector should near shade during the hours of sunshine and thus favors a habitat where bush abounds or where the lower branches of trees provide a resting place. Acacia bush and savannah are thus frequently the favored habitats of the tsetse.

As with diseases mentioned earlier, the problem of cattle movement is considerable, and in nomadic areas, treks for water and grass into or near the fly belt often result in the introduction of disease into otherwise uninfected tsetse country.

7.5.1.4. East Coast Fever

This is a tick-borne disease which causes extremely heavy calf losses, considerable stunting and loss of production. The distribution of East Coast

Fever (E.C.F.) is strongly governed by the ecological requirements of the vector: the tick. Ticks require a rainfall above 25 inches and an altitude below 7,000 feet in areas of fairly long grass (Figure 7.7).

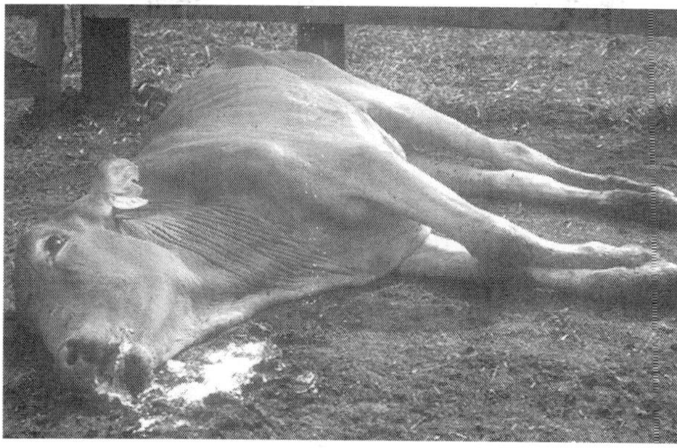

Figure 7.7: A beef cattle suffering from East Coast Fever (*Source:* http://vetbook.org/wiki/cow/index.php?title=East_coast_fever).

As there is no cure for E.C.F., control of the disease is achieved by eradicating the vector. At present E.C.F. has a more serious effect on the production of slaughter cattle than all other diseases combined.

7.5.1.5. Foot and Mouth Disease

This disease has little impact in most parts of the world. The unimproved nature of most of the livestock results in the disease having only a very slight effect on production or productivity. Nevertheless, this disease as much as any other is responsible for considerable reduction in meat on the world market and until Foot and Mouth is eradicated there is little prospect of export to better-priced outlets (Figure 7.8).

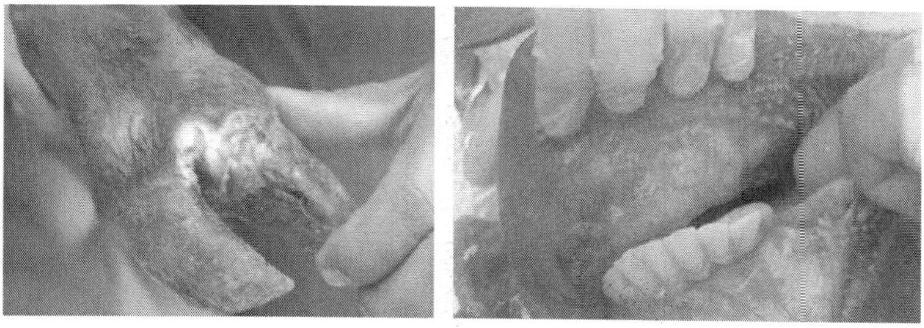

Figure 7.8: Foot and mouth disease (*Source:* https://www.agric.wa.gov.au/livestock-biosecurity/foot-and-mouth-disease-recognize-and-report-signs).

7.5.1.6. Other Disease

Parasitic infestations such as *Cystioercus bovis* and liver-fluke are very common and lead to heavy condemnations of slaughtered stock. The presence of *Cysticerous bovis*, an intermediate stage of the tapeworm *Taenia saginata*, results from the close association of cattle and man and is thus most commonly found in cattle originating from the pastoralist areas where the symbiosis of man and beast is closest. Liver fluke and related parasites are often contacted at water holes and points where cattle gather in large numbers to drink. Transmission is rapid and losses through condemnation are often high.

7.5.2. Social Attitudes towards Cattle

Social attitudes towards beef cattle farming in every country around the world are a considerable influence over the regional characteristics of slaughter stock production and sale. Attitudes extend not only to the stock themselves but also over the areas grazed and the water drunk.

Although the place of beef cattle in the economy of any country may be a reflection of the poor climatic prospects for reliable crop agriculture, this is by no means a certainty. It is impossible to be deterministic about social attitudes, their origins and adoption, though it is easy to see how pressures from cultivation gradually push the more versatile activity, pastoralism into the less clement areas. At present social attitudes are important not only in influencing the level of production, the use of resources, the pattern of sale and the response to advice in many areas but also act very often as a considerable barrier to progress (Gebremedhin, 2002).

Worldwide, beef cattle farmers can be classified into three main socio-cultural groups with regard to stock keeping. These are:

- The Pastoralist.
- Pastoralist-Cultivator.
- The Cultivator.

7.5.2.1. The Pastoralists

These are people to whom beef cattle form the basis of subsistence, exchange, prestige and culture. Some tribes may involve themselves only with cattle, shunning cultivation. Others may grow a few grain crops.

7.5.2.2. Pastoralist-Cultivators

In this type of group, cattle have lost their pivotal place in the society and culture but still play a huge role in prestige and the acquisition of a bride. However, the tribes in this category are well within a cash economy and regard their cattle as having a cash value as well as a value within the tribal standards.

7.5.2.3. The Cultivator

This group has little cultural or social interest in beef cattle. The cultivators regard beef cattle as economic rather than a social asset.

7.5.3. Water Resources

It has been estimated that most indigenous beef cattle normally require five to ten gallons of water per day). Whether or not this is obtained under natural conditions depends in any area on:

- The total rainfall.
- The seasonality of the rainfall.
- The incidence of swamp, rivers, lakes and sand beds.

Present natural water supplies are inadequate and quite unsatisfactory for beef cattle farming in most countries. Most of the drier countries with an unreliable rainfall can lead to heavy losses of cattle. As important as the mean annual total and reliability of the rainfall is the factor of seasonality (Abington, 1992).

During the long dry season nearly all-available surface water dries up and this partly necessitates a pattern of transhumance.

Natural water supplies are a considerable barrier to production improvement in the drier areas and improved stock cannot be kept on traditional sources of water. Most water becomes heavily polluted with parasites in the dry season as an increasing number of animals drink from the same source. The search for water in the dry season often involves treks in the sun and this walks off a large part of the weight gain which is particularly art, important if the animals fee to be raised primarily for beef (Sutherland, Wilson, & Zagata, 2014).

The amount of dry matter consumed and, with it, the amount of weight gain are directly related to the quantity of water available and thus water becomes a major factor in the efficient utilization of pasture resources.

Disease control work is largely nullified if cattle have to trek daily through tick-infested areas in search of water. The concentration of stock around a very limited number of watering points also leads to grazing out of the peripheral areas, the limiting of the grazing during periods of waiting and serious erosion (Figure 7.9).

Figure 7.9: Cattle trekking for long distance in search of water (*Source:* http://www.utoonidevelopment.org/no-need-to-cattle-queue/).

In the wetter areas, natural water supplies are unsatisfactory as they are often in permanently swampy areas which may encourage foot rot. The sources of water for cattle are often the same as those for man and this creates the ideal breeding condition for tapeworm, which depends on man as a host during part of the life-cycle.

7.5.4. Environmental Factors that Affect Grazing Behavior

7.5.4.1. Temperature

Studies have shown that temperature difference during day affects the grazing behaviors of most beef cattle. For instance, Bonsma, Schaltz & Badenhorst (1940) in South Africa, showed that Hereford cattle in sunny conditions with temperatures of 39 °C reduced daytime grazing to 79% of 7.9 h average grazing time while Brahmans reduced to 89% of 8.5 h average grazing time.

However, some researchers have shown that indigenous beef cattle in the tropics are slightly influenced by a rise in environmental temperatures on time of daylight grazing. For instance, Smith (1959) showed that temperatures had no influence upon Zebu cattle grazing time from 6.00 to 20.00 h. In cool months when maximum temperatures were 21.9 °C, grazing time for the overstocked group of cattle averaged 8.3 h per day between 6.00 to 20.00 h local time and 7.9 h for the mildly stocked. In hot dry months when maximum temperatures varied between 26.1 °C to 36.6 °C, the overstocked cattle grazed 9.8 h while the other group spent an average of 8.6 h grazing.

The grazing time distribution pattern and intensity has been shown, however, to be affected by rising environmental temperatures (Kyomo, Hutchson and Salehe, 1972).

Studies have also shown that temperature has some influence on idle time of the beef cattle. In this context, idle time refers to time cows spent lying resting, standing resting, walking and drinking. For instance, under higher environmental temperatures most exotic beef cattle spent more time lying and resting. This is because higher temperatures and the greater number of hours of direct sunlight causes an increased heat burden on the grazing cattle resulting in a longer time spent idling in lying position.

It has also been shown that watering behavior of beef cattle is influenced by environmental temperatures. The higher the temperatures the more time animals spend drinking water.

7.5.4.2. Sunlight/Darkness

Two major grazing peaks occur each day throughout each season during the year. The first peak of intensive grazing starts shortly before or after sunrise and normally lasts for 2 to 3 h. Another peak normally commence in late afternoon or early evening which extends to shortly after sunset.

It has also noted that 100% grazing usually take place during daylight in mid-summer partly because of extended daylight to 18.83 h. In autumn there is a tendency for cattle to graze at midnight. The total grazing time is normally 32% of the 24 h total time steers are on pasture.

It is generally accepted that rumination occurs mostly during night hours. Wilson (1961) reported that 63 to 76% of the total 7 h daily ruminating time took place at night. Goldson (1963) noted 67% of the 6.5 h total ruminating time taking place at night.

It has been shown that most cattle are normally inactive at night and usually do little walking, but spend most of the time lying while ruminating and resting alternatingly.

7.5.5. Pasture Quality and Quantity

Grazing time is described as time an animal spends in the act of eating herbage including short periods of walking while selecting grass on pasture field.

Most cattle spend 30% more grazing time on poorer pastures than those on better pastures in terms of quality and quantity. However, Johnstone-Wallace (1944) showed that grazing time was not related to pasture quality or quantity. They showed that beef cows spent 7 to 8 h grazing per day and as grazing progressed better pasture was progressively selected thus forage quality deteriorated and quantity decreased but time spent grazing by cows did not change.

It has also been reported that steers on a pasture of comparably more TDN (Total Digestible Nutrients) spent less grazing time than steers on a pasture of comparably less TDN.

Several studies, carried out in East Africa and elsewhere in the tropics concerning grazing time of Zebu (Bos indicus) cattle, have shown that grazing time increases with poorer quality and limited quantity of pasture. For instance, Harker et al (1956) in Uganda studied habits of ten adult cows for four periods of 48 h each in different months of the year. He showed that free grazing cows throughout day and night increased their grazing time during dry season when pasture quantity was lower and quality poorer. In June, when there was relatively good quantity of herbage, cows grazed an average of 9.7 h per day (24 h) while in mid-January, towards end of dry season, when pasture was limited, cows grazed for an average of 10.4 h per day.

Ruminating is the second important time taking activity by cattle on pasture. Rumination is closely related to feeding. Rumination has been found to relate to type of food ingested. Feeds of higher digestibility have been demonstrated to cause shorter times of rumination in cattle (Lofgreen et al , 1957). Gordon (1958) showed that diets of concentrates only reduced time of rumination but diet which includes roughages such as hay increase ruminating time. However, no relationship exists between quantity of hay fed and the time of ruminating.

Watering behavior of cattle may be influenced by type of feed and quality. It has been reported that Hereford steers spent 0.2 h per day (24 h) on drinking water on abundant alfalfa pasture but spent average of 0.1 h on scant alfalfa pasture. At times when pasture is wet by rainwater or dew cattle tend to reduce drinking water habit because partly or all the water intake requirement is met from the wet grass. In a coarse drier pasture, animals tend to spend more time on watering.

7.5.6. Breed Heat Tolerance Differences

Several studies show that most indigenous beef cattle are more adapted to hot environment than the exotic beef cattle. Indigenous beef cattle normally have a thinner skin thickness compared to exotic breeds. The combination of thinner skin thickness and larger surface area, as a result of the dewlap in indigenous beef cattle provide a means of dissipating excessive body heat through radiation, convection and conduction to the surrounding.

Cattle breeds differ in their ability to sweat and that high sweat gland volume is related to high heat tolerance (Steinhauf, 1977). Evaporation of sweat is influenced by lowering skin temperatures. Taneja (1960) also showed that indigenous beef cattle have more sweat glands than exotic beef cattle especially on the shoulder area.

The coats of breeds of cattle are closely related to heat tolerance (Bianca, 1959). Indigenous beef cattle have a smooth hairy coat and the hair is straight which facilitates cutaneous evaporation readily. The hair emerges from the-better developed sweat and sebaceous glands as compared to those of exotic breeds. The sebaceous glands secrete more sebum that prevents the drying of superficial layers of the skin and may afford some advantages in reflecting solar radiation (Bonsma, 1949). The coat of indigenous beef cattle has two types of hair, modulated and non-modulated. The non-modulated is coat's inner thin curly hair which forms a mat on the skin and retains heat from the body. It is for this reason that the animals have difficulties to dispose of excessive heat effectively through radiation and evaporation in hot environment (Taneja, 1959).

7.5.7. Influence of Night Kraaling on Grazing

It has been shown that night paddocking has significant influence on grazing time of adult Zebu cattle especially in dry season. In times of adequate pasture there is no difference of grazing time between free grazed cattle and the night paddocked ones. It has generally been shown that under milder climatic conditions kraaling cattle at night is not important but in areas and conditions of grass shortage, kraaling cattle at night limit grazing time (Harker, 1956).

However, there is no significant difference of live weight gains between free grazed cattle and the night kraaled cattle at any period of the year. The kraaled cattle during pasturing graze more or less continuously. Only in wet season do the animals stop grazing and rest. Probably, kraaled cattle normally attempt to overcome limited grazing time by grazing intensively and by postponing ruminating and resting until they had been kraaled in the evening.

8
CHAPTER

SUSTAINABILITY OF POULTRY, RABBIT, AND FISH FARMING

CONTENTS

8.1. Poultry Farming 188
8.2. Rabbit Farming 194
8.3. Fish Farming 203

8.1. POULTRY FARMING

8.1.1. Introduction

Globally, poultry products are among the most plentiful sources of animal protein for all classes of people. Compared with other types of livestock products such as cow's milk and meat, poultry eggs and meat are the most economically produced animal protein. Biologically, poultry products are also ranked highest (Leitch, Isabella, and Godden, n.d.).

In most places around the world, poultry is reared either for the supply of animal protein to the family in rural areas or for the supply of protein to the urban population and other concentrations of money-earning communities. Therefore, there is the need to improve on the supply of poultry products to cater for the ever-expanding population is much more felt now than ever before.

To have a productive and sustainable poultry rearing business, poultry farmers must ensure the birds are provided with well-ventilated clean housing and balanced diets. The birds should be provided with a comfortable environment so that they achieve their maximum genetic potential for production because, with the right kind of stock and suitable housing and food supply, production of eggs is likely to be in direct proportion to the comfort of the birds (Abington, 1992).

8.1.2. Poultry Housing Systems

There are three types of intensive poultry housing systems that can be used for sustainable poultry production (Table 8.1). These include:

- Deep litter,
- Wire floor, and
- Cage systems.

Table 8.1: Types of Poultry Housing Systems Used in Intensive Farming

Poultry housing system	Characteristics
The deep litter 	This system has the advantage of being more flexible than the other systems. It is also reported that birds housed under this system feel comfortable. Maintenance costs are low, but there is a high risk of bacterial diseases and internal parasites. There is a tendency of collecting dirty eggs, and culling of the low producers is difficult.
The wire floor 	In this system there is better control of bacterial diseases and internal parasites. There is usually high humidity in the houses, birds appear nervous, feather pecking is high, and there is a high rate of egg breakages.
The cage system 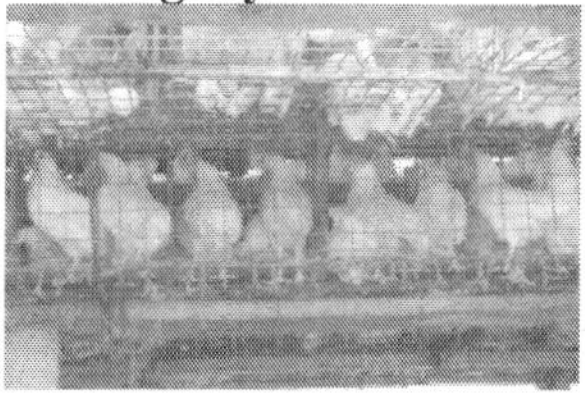	This system accommodates more birds per given floor area than the other systems. The system also eliminates bacterial diseases and internal parasites. The problem associated with it is that capital requirement is high, removal of manure is difficult, and there is high labor requirement.

Source: http://agritech.tnau.ac.in/animal_husbandry/ani_chik_poultry%20rearing.html.

It has been noted that most modern poultry commercial poultry farmers normally retain a preference for one housing system or the other without sufficient experimental support. Reports on the effect of housing systems on poultry production have been contradictory.

It has been shown that most poultry birds usually feel comfortable in a deep litter housing system than any other system. However, it is worth noting that this type of housing system is susceptible to high incidences of bacterial diseases (Gibon & Mihina, 2003).

Although most birds normally appear nervous most of the time on wire floor housing system, this type of system is better when it comes to diseases control. Caged birds on the other hand usually have higher feed conversion efficiency but they suffer from fly infestation because it is difficult to control flies in this system. With respect to egg production caged birds also perform better than those housed in other systems (Leitch, Isabella, and Godden).

Therefore, housing of poultry for egg or meat production is a significant task because it helps in keeping environmental conditions under the control of the operator. It also helps to ensure maximum use of feed energy for productive purposes rather than overcoming the effects of unfavorable weather.

Every poultry farmer must have certain production goals in his/her poultry farm. Some of the realistic poultry production goals include:

- Having production of at least 240 eggs per hen annually.
- Feed conversion of less than 2.04 kg per dozen eggs.
- A laying house mortality of less than 10% per year.
- Having production of 95% or more marketable eggs.
- Having production of 75% or more large grade eggs.
- Having production of on farm egg breakage of less than 2%.

It is important to note that most of these goals are customarily affected by the housing systems adopted. Studies have shown that most of the qualities of the eggs production (such as egg size, eggshell thickness, blood spots in eggs, feed efficiency, live weight gains, and mortality) are all affected by type of housing system that a farmer adopts.

8.1.2.1. Rate of Egg Production

The rate of production is the total number of eggs produced regardless of their quality. Research has shown that the performance of poultry birds under cage housing system is always better that those under deep litter housing system. It has been noted that caged poultry birds on average produce 3% more eggs compared to those that are housed under the deep litter system.

8.1.2.2. Egg Size

This refers to the average weight of an egg usually expressed in grams. It is usually affected by the energy and protein levels in a feed. A decline of energy and protein levels in poultry feeds will most often lead to a drop in egg size. Therefore any system which might affects feed intake by the birds will cause a reduction in egg size.

8.1.2.3. Shell Thickness

The thickness of an egg is another very common trait that is used to determine the quality of eggs laid by poultry birds. It is thought that measurement of eggshell thickness can provide vital information in relation to the liability of an egg to crack, and possibly also in relation to hatchability. However, studies have shown that thinner shells are not necessarily weaker than thicker shells. This is because factors affecting the strength of eggshells are heredity, age, health, season, production rate, diet and environment. But in the absence of equipment for measuring eggshell strength, thickness can be used to determine shell strength.

8.1.2.4. Blood Spots

The incidence of blood spots in eggs normally cause great losses in marketable eggs as well as unfavorable reactions from consumers. It is commonly assumed that 1–6% of chicken eggs randomly selected will contain blood spots. Studies have shown that most blood spots on eggs are caused by rupturing of blood vessels in the follicle wall or hemorrhages that occur before ovulation.

The extent of blood spot incidence is influenced by different factors. These include:

- Differences in heredity
- Nutrition
- Age
- Season
- Management, and
- Stress

The tendency of birds to lay eggs that contain some blood spots has a relatively high heritability of 30%. However, Parkhurst, Carmen, & Mountney (2012) shown that deficiency of vitamin A can also increase the incidence of blood spots as will high levels of protein.

8.1.2.5. Feed Efficiency

Food efficiency is the maximum use of a given amount of feed per unit of animal product produced. When compared with the total number of egg produced, it is usually the amount of feed consumed per given weight of eggs produced. In poultry production it is normally impacted by the system of management employed.

For instance, research has shown that caged birds usually have a better feed efficiency than those kept on deep litter. This means that most caged birds use feed more efficiently than layers on deep litter.

8.1.2.6. Live-Weight Gain

Although most poultry farmers are usually concerned about the number of eggs of good quality that the birds produce, live weight gains of the layers are also significant. They give an indication on how well the birds are kept, and the heavier birds fetch more money at culling time.

8.1.2.7. Mortality

Birds' mortality in any type of poultry production normally shows a direct financial loss to the poultry producer. The type of a housing system may significantly affect the mortality rate of the birds. For instance, studies have shown that birds housed under cage system normally have a lower mortality rate than those housed under deep litter.

8.1.3. Marketing of Poultry Products

According to OECD (2003), marketing is defined as the performance of all business activities involved in the flow of goods and services from the point of initial (agricultural) production until they are in the lianas of ultimate consumers. Different people view marketing in a different way from their own points of view. For example, consumers are interested in receiving what they want at the lowest possible price. On the other hand, farmers are interested in gaining the highest possible returns from the sale of their produce. The various traders engaged in doing a variety of marketing tasks are interested in maximum profitability of their particular business operations. The study of the marketing of poultry product therefore is of interest if it attempts to throw some light on how such conflicts are compromised among the various groups seeking different economic goals.

High marketing margins may arise from collusion or from inadequate allocation of resources, either at the grass-roots level or at the national level. For instance, poultry farmers may not necessarily produce the types, quality or seasonal amounts of products which consumers want and distributors may be inefficient and therefore cause the high margins. Furthermore, marketing services of poultry products affect the level of demand in that an efficient service provides the necessary intelligence network and a good transport system which finally ends in reducing costs end helping to maintain or increase effective demand.

Therefore, there is the need to investigate the efficiency of the marketing mechanism within any agricultural sector for example poultry if only because marketing has an important bearing on agricultural production.

Improving the marketing system is essential, in order to provide farmers with, greater incentives to adopt new technologies. All of the improvements available to increase farm output involve added costs; for instance:

- Improved breeds of poultry
- Better housing facilities
- High quality feeds and
- Improved feeding methods

The existence of an efficient and flexible agricultural marketing system makes it more readily possible to achieve a smooth transformation of the agricultural sector in line with national development strategies.

8.1.3.1. Demand for Poultry Products

Demand for poultry products has always been high globally. This increase in demand would be influenced by the cross elasticity of poultry produce which their close substitutes (FAO, 1993). Distribution of income is also likely to become less unequal due to various government policies directed to this end and several international developments which supplement the government efforts. For instance, several donors' countries such as the United States are orientating aid programs especially towards improving the receipts of the low income groups. This redistribution of income towards equality would therefore be expected to result in an increased per capita consumption of poultry produce among others (FAO, 1993).

Another determinant of demand for poultry produce is population and population growth rate. There is every indication that the world's population will continue to increase and to this extent, therefore, there will be an increasing demand for poultry produce.

Another factor which may influence the demand for poultry meat especially is the level of poultry meat prices vis-a-vis other livestock meats, and more so beef price levels. Studies in the United States of America and Western Europe have shown that reduction in prices of chicken meat and eggs have resulted in a proportionate increase in consumption of these products as the incomes of consumers rise.

8.2. RABBIT FARMING

8.2.1. Introduction

The domestic rabbit belongs to the vertebrate family *Laporidae* (Order: *Lagomorpha*), and is descended from wild rabbits of Western Europe and North West Africa.

Domestic rabbit play an important role in many households worldwide. They are a good source of meat. Studies have shown that most people are currently rearing rabbit in large-scale because they are efficient producers of meat under extremely varying conditions (Cheeke, 1986). In other words, rabbits are adaptable to almost all climatic, feeding and holding conditions.

Compared to most of the other mono-gastric animals, studies have shown that rabbits can endure on the low-energy high-fiber feed from weaning to slaughter and they can use diets with up to 30% fiber. Their protein requirement is also known to be low. For instance, although the amino acids requirements of rabbits seem to be similar to that of chicken, their protein requirement at 12–16% is substantially less than that of chicken at 23% (Cheeke, 1986).

Furthermore, rabbits are efficient converters of feed into body protein and their growth rates are comparable to those of chicken. In comparing rabbits with other livestock raised for meat it has been noted that their energy requirement per unit of growth is below that determined for sheep and cattle, that slaughter weight is reached relatively quickly, that annual meat production of the dam is greater than for any other meat producing herbivore and that protein production per unit area is higher than in lambs.

Still, the quality of rabbit meat may be superior to that of most other animals. Rabbits have an additional advantage because of their size. Being small and therefore of low individual value, rabbits have limited risk in investment and can easily be afforded by low-income households with regard to purchase price and care. Additionally, the small size of carcass does not present the storage problems associated with other livestock (Cheeke, 1986).

Even though in many developing countries the human animal protein needs are not adequately met, the potential role of rabbit meat as a source of protein to improve human nutrition remains to be exploited. The rabbit owes its suitability for meat production in these countries to its ability to effectively convert forage feed into animal protein (Leitch, Isabella, and Godden). Since these animals can efficiently subsist on roughage based diets, a farmer on a small holding can make better use of his limited land by feeding crop-residues to such animals.

With the rapidly increasing human population, especially in developing countries, the trend towards more intensive and semi-intensive rearing of domestic animals has to be extended to rabbit production. To effect these transformations, research should be aimed at identifying ways of maximizing the utility of rabbits. Such studies should be aimed at selecting breeds of rabbits suited to various management practices, enhancing their complementarity with, and integration into crop cultivation, and reducing their competition with man for vegetable and other human food resources.

8.2.2. Challenges in Rabbit Farming

Very few studies on domestic rabbit meat production have been published, especially in developing countries. Therefore, there is lack of information on:

- Breed characteristics.
- Genetic potential.
- Performance levels.
- Environmental tolerance under conditions.

Such information would be used as a basis for identifying strains with high reproductive rates, early maturity and low death rates. Little is also known in most countries about the nutrition and management factors influencing rabbit production in terms of growth rate, feed efficiency and feeding economics (Abington, 1992).

Apart from the value of tropical crop residues for rabbit, meat production has not been fully assessed. Studies have shown that considerable breed differences exist in the growth potential of rabbits. Of all the factors of animal production, feed accounts for the biggest fraction of total cost, and its optimum utilization is influenced by the breed of the animal used. It is thus necessary that rabbit production programs should combine the cheapest feeding system with the most productive animal breed.

It is believed that small-scale farmers could usefully and profitably integrate small-scale rabbit production into their activities by using crop-residues and forages in combination with minimal supplementation with commercial concentrates. In Africa for instance, substantial quantities of crop residues that could be used in rations for rabbits at present go to waste. Intensified crop production in which legume crops are used in fallow areas would further increase suitable feed for rabbit stock (Cheeke, 1986).

8.2.3. Global Rabbit Meat Production and Consumption

Research indicates that annual rabbit meat production globally is in excess of 1 million tons (Table 8.2).

Table 8.2: Showing Some of the Largest Rabbit Producers in the World As of 1990

Countries	Production (tons/annum)
Soviet Union	240,000
France	180,000
Italy	150,000
Spain	120,000
China	With more than 40,000
Hungary	40,000
Poland	35,000
Portugal and German Democratic Republic	20,000 each)
West Germany, Great Britain and USA	each with 15,000

Source: http://www.rabbitadvocacy.com/pdf_files/Rabbit%20 Industry%20Production%20EU.pdf.

The highest annual per capita consumption of rabbit meat is in Malta at 8.0 kg. The estimated per capita consumption for France is 6.0 kg, Italy and Spain 2–3.0 kg, Switzerland 1.0 kg and West Germany 0.2–0.3 kg.

World rabbit production figures however plainly show that apart from France and Italy, large-scale rabbit production is associated with the relatively less advanced economies of Western Europe, Communist Europe and China. It is further evident that rabbits are still not significant as sources of meat in developing countries (Cheeke, 1986).

8.2.4. Types of Rabbit Production Systems

There are three types of rabbit production systems. They are mainly distinguished depending on the level of production inputs purchased by the producer and the degree of commercialization. These are:

- Extensive system.
- Semi-intensive system.
- Intensive system.

The extensive system is the most widely practiced form of rabbit keeping in most countries. It is normally characterized by the keeping of one or two local rabbits for home consumption. The feed consists of garden weeds (Figure 8.1).

Figure 8.1: An extensive rabbit rearing system (*Source:* http://www.monitor.co.ug/Magazines/Farming/Rabbit-farming-prospect-food-nutrition-income/689860–2988572-vii7lu/index.html).

In semi-intensive system, rabbits are normally kept for both market and home consumption. In this system, hutches with wire floors are used for accommodation and animals are fed a combination of forages and concentrates (Figure 8.2).

Figure 8.2: Illustrating a semi-intensive rabbit rearing system (*Source:* http://smallbusiness-jambuabang76.blogspot.co.ke/2014/02/commercial-rabbit-farming-business.html).

The intensive system involves keeping rabbits in wire cages and feeding them commercial concentrates. This system is common in developed countries, but in most developing countries, intensive rabbit keeping is rare (Figure 8.3).

Figure 8.3: An intensive rabbit rearing system (*Source:* https://www.alibaba.com/product-detail/commercial-rabbit-cage_1180949874.html).

For increased rabbit production in developing countries, studies must be conducted to adapt the various production systems in order to exploit the available resources.

8.2.5. Reproduction and Development of the Rabbits

The expression of live weight gain and the growth rate of the main tissues depend on the breeds' biological characteristics and on the production factors such as feeding. The age at first breeding of female rabbits is dependent on the nutrition level and the corresponding body development. Thus there is a range in age at first breeding from 4 months with a diet of pelleted concentrates, to 8–10 months on forage feed (FAO, 1997).

Gestation is dependent on the proper function of corpora lutea and lasts 31–32 days. After the 18th day of gestation, the doe shows signs of preparing for kindling by nest building. The young are born naked and blind and remain dependent on the does' care until they are about 3 weeks of age. At this age they begin to consume solid feed.

On average, a well-maintained doe can produce 7–9 litters per year consisting of 8–9 young per litter. The progeny is ready for slaughter at 10–12 weeks of age under average conditions of management. It has been noted that as bone tissue in rabbits develops first followed by muscle and fat a poor feed will slow overall muscle gain as most nourishment will be used for skeletal development.

Most of the fastest growing animals in a population will have the best carcass composition (muscle/bone ratio, fat percentage) at slaughter age or weight (FAO, 1997).

Pre-weaning growth performance is one of the important traits for rabbit meat production. The trait has a direct influence on weaning weight which can in turn be employed to determine meat production per doe per year. The pre-weaning growth rate and weaning weight in the rabbit are determined by several factors. These include:

- Breed size.
- Weight at birth.
- Litter size.
- Maternal effects such as nutrition and milk production.

Young rabbits' growth rate is highly influenced by the size of the adult and weight where there has been no marked dietary deficiency. Adult size is largely determined by breed. For example, under similar conditions of management, young small Himalayan rabbits (adult weight of 2.5 kg) normally have a slower growth rate compared to the New Zealand white breed (adult weight 4 kg) (FAO, 1997).

Studies have shown that the genotype of rabbits normally affect pre-weaning growth. In rabbit production, the tendency is for large breeds to weigh more at weaning than light ones.

Birth weight is also an important factor of growth performance, and weight at weaning. Therefore, rabbit birth weight has been shown to be positively correlated with weaning weight genetically and heavier litters at birth normally have correspondingly higher weights at weaning.

However, studies have shown that weaning weight in rabbits is more a reflection of the mothering ability of the doe than that of the potential of the young since it is strongly influenced by maternal effects such as nutrition and milk production. It can therefore be concluded that a large fraction of the variation in growth rate before weaning can be accounted for by the differences in milk yield of the dams. Furthermore, it is known that litter size influences pre-weaning growth rate by putting a limit to the amount of milk available to each kit (FAO, 1997).

8.2.6. Rabbit Nutrition

Rabbit nutritional needs entail an adequate and balanced supply of amino acids, energy, minerals and vitamins necessary for the growth and development of the animal. The content of protein and energy in the diet normally has a significant influence on live weight gain of the rabbits. The proportions of the rabbits' body parts can also be manipulated by dietary means (Cheeke, 1986).

Studies have also shown that nutrition has an influence on the sexual development of the rabbit. This means differences in nutrition will cause a variation in the age at first breeding of female rabbits from 4 months with a diet of pelleted concentrates to 8–10 months on forage feed.

Therefore, for rapid and efficient growth, rabbit farmers must ensure that the diet of the rabbits contain at least 17–18% crude protein and an energy food source equivalent to 3600–3700 Kcal/Kg in the diet. In a study, Ouhayoun et al. (1981) used two diets of 13 and 16% crude protein and similar energy value to feed weaner rabbits of the California breed. They reported that animals fed 13% crude protein took 6 days longer to reach 2.2 Kg body weight although efficiency of feed utilization was similar between groups when calculated over the duration of the experiment.

It has also been noted that slaughter yield improves with age and that animals with a high growth rate, receiving more balanced feed generally have a better carcass yield. The level of protein in the diet also influences dressing percentage. Dressing percentage is reported to increase with increasing protein levels. After feeding rabbits on diets containing 18.9 and 16.4% crude protein and 12.03 and 18.21% crude fiber, respectively, Zaragoza & Vallejo (1982) reported dressing percentages of 59.9 and 57.4 from the two groups of animals, respectively.

Variation of source, amounts and proportions of crude fiber in the rabbit diet may also influence the growth and meat production of these animals. By varying the proportions of sunflower husks in the diet, Gippert et al. (1985) obtained rations with crude fiber contents varying from 10 to 16%. The greatest body weight gains were recorded for rabbits fed rations containing between 11.6 and 13.8% crude fiber.

The efficiency of feed utilization is improved by increasing the percentage of fiber in the diet. For instance rabbits fed diets containing 28, 54 and 74% alfalfa in the diet yielded dressed weights of 46.7, 48.7 and 46.8%, respectively.

Research shows that feeding rabbits with increased quantities of crude plant materials such as hay may result in reduced carcass yields. Furthermore, while there is a tendency for dressing percentage to increase with increasing protein levels, the visceral weights increase with increasing fiber levels in the diet. It is therefore clearly evident that nutrition is a vital component of rabbit meat production.

8.2.6.1. Forages in Rabbit Production

As livestock production gains significance, the cost of commercial feeds continues to rise, especially in developing countries. Forages play a crucial role in substituting most nutritional requirements found in commercial cereal feeds. Economically, forages are cheap and plentiful throughout the year in most parts of the world, and can be preserved for the dry period in the low rainfall areas (Abington, 1992).

The potential of forages is particularly significant for rabbits as they are capable of digesting leaf proteins effectively. Research has shown that young rabbits normally consume sufficient quantities of low energy high fiber diets to grow from weaning to slaughter at 2.2 kg live weight. In a practical low input system, such diets would have to support growth from conception to slaughter.

Studies have revealed that most rabbit breeds can produce five times more meat than cattle from the same amount of an alfalfa based diet. High fiber material which is usually a constraint in poultry diets is beneficial to rabbits for preventing enteritis and, well-preserved roughage reduces digestive problems and improves palatability of the diet (Abington, 1992). Moreover, it has been noted that although rabbits are not able to obtain as much energy as ruminants, from forage diets they are able to consume these in quantities sufficient to meet their energy requirements. However, the nutrient content of forages varies significantly among species, among cultivars within species and even among stages of growth in the same cultivar. This variation in nutritive values can be evened out by some form of supplementation.

Where rabbits are fed on concentrate feed, inclusion of forage material is highly valuable. For instance, Fayeye & Ayorinde (2010) reported a reduction in feeding costs without affecting performance by adding hay to the diet to reduce the intake of concentrate feed. In these studies, a combination of concentrates with lucerne or grass hay resulted in daily body weight of 40.9 and 32.8 grams respectively, compared to 36.4 grams for animals fed on concentrate pellets alone.

The level at which a forage should be included in the diet depends on its quality. The type and quality of forage used are therefore important factors to consider in rabbit feeding.

8.2.6.2. Commercial Concentrate in Rabbit Production

Forage consumption alone is usually not sufficient to fully exploit the rabbit's potential for growth. This is due to the fact that most domestic rabbits lack facilities for selective ingestion of plants and plant parts with high nutrient density. Besides, the capacity of digestive organs as a ratio of body weight is very low for the medium weight breeds of rabbits which are chiefly used for meat production. This means that the capacity for a rabbit to compensate for a low concentration of nutrients in the fodder by increasing intake is limited.

Ghosh (2010) shown that the intake limiting effect caused by insufficient protein content in forages can be compensated by minimal supplementation. Young animals, however, require a higher proportion of concentrate feed in their ration as their ability to compensate for a lower concentration of nutrients in the feed by increasing intake is even less than in the adult.

When concentrate supplementation is constant, growth performance is influenced by the forage material used for feeding the rabbits (Ghosh, 2010).

Because of economic and other management considerations, it has been noted that the amount of commercial concentrate feed a rabbit farmer is recommended to give to his animals cannot be based solely on results from nutrient balance studies. While the optimum level of feeding has to be defined in economic terms, it is also necessary to limit the amount of commercial concentrate in the ration of both fattening and breeding animals. This is because the high feed intake obtained for pelleted concentrate, apart from increasing the cost of production causes extensive fat formation which has a detrimental effect on carcass quality and the reproductive performance of breeding animals (Ghosh, 2010). There is therefore a clear need to combine forage and concentrate components in rabbit diets in such a way that costs are minimized and performance maximized.

8.3. FISH FARMING

8.3.1. Introduction

Globally, fish farming practices have significantly grown over the past half-century to around 52.2 million tons in 2008 worth US$ 98.5 billion and in turn representing about half of the world's fish food supply. According to FAO (2014), food fish supply has been increasing in the last five years at an annual average rate of 3.2%, surpassing the world population growth of 1.6%.

In sub-Saharan Africa, small-scale fish farming is a recent activity (FAO, 2004). However, it continues to steadily grow with millions of poor people relying heavily on fishing and farming to earn their livelihood and feed their families.

Most Sub-Saharan African countries rely heavily on fish as a significant source of animal protein (FAO, 2004). Some of key roles played by fish farming activities include:

- Sustainable food production.
- Contributes towards economic development.
- Contributes to food security.

8.3.2. Types of Fish Farming Methods

8.3.2.1. Extensive Fish Farming

It is practiced in pond or along ditches. Most fish farmers practicing this method normally use locally available raw materials to feed their fish. The feeds used in this type of fish farming method vary depending on availability.

Studies have shown that some fish farmers in Africa use feed supplements such as maize bran, rice, and kitchen leftovers to feed their cultured fish stocks in ponds (Figure 8.4). Some of the challenges experienced with this type of fish farming include:

- Lack of appropriate technologies.
- Lack of proper breeding stocks.
- Lack of market.
- Lack of credits.
- Lack of viable fish farming information.

Figure 8.4: An intensive fish farming system (*Source:* http://www.eco-business.com/news/coordinated-efforts-aquaculture-needed-meet-global-demand/).

8.3.2.2. Intensive Fish Farming

This type of fish farming method is normally practiced in ponds, cages, tanks, and raceways. Water is obtained from nearby streams and rivers. This fish farming system is normally practiced by companies who produce fish for export. There are several factors used in characterization of this type of fish farming (Figure 8.5). Some of them include:

- The type of fish farming facility (for instance, raceway, cages, or ponds).
- The ration between total farm area and facility areas.

- Water used.
- The use of inorganic fertilizers.
- Fish species.
- Stocking densities.
- Labor requirements among others.

Figure 8.5: Intensive fish farming system (*Source:* http://www.aquabait.com/about_aquabait_marine_worm_aquaculture.phtml).

8.3.2.3. Integrated Fish Farming

This fish farming method involves integration of fish farming together with crop farming. In other words, various crops are normally planted around the fish ponds. It has been noted that integrated fish farm ponds usually enable farmers to spread the risks due to crop failure on seasonal croplands by retaining water on the farm.

Research has also shown that fish productivity of integrated small farms is usually higher compared to non-integrated fish farms. This translates to more profits in integrated fish farming systems (Figure 8.6).

Figure 8.6 Integrated fish farming system (*Source:* https://gdblogs.shu.ac.uk/b4017761/2016/10/19/aquaculture-methods-rice-fish-cultureintegrated-fish-farming/).

8.3.3. Sustainability of Small Scale Fish Farming

Sustainability is one of the most significant challenges in today's times. This is because most donors funded projects die as soon as the donor pulls out. According to Western Cape Education Department, "sustainability as a concept of development is one that meets the need of the present without compromising on the future generation to meet their needs." Sustainability is very critical and must be addressed as a requirement in any projects during planning and design stages (Reza, 2012). Sustainability involves balancing or harmonizing social, environmental, and economic interests considering the full life cycle of any project. Economic sustainability has been defined as effectively and efficiently using available resources to ensure that the business continues to return profit over several years. According to Dillard, Dujon, & King (2008), there are many ways to look at economic sustainability nevertheless: Economic sustainability needs current economic activities which do not jeopardies or burden the finances of the future generation.

For small-scale fish farming projects to be economically sustainable, fish and fisheries products should be produced at a competitive cost with other animal-protein sources and sold at a reasonable profit. Other sustainability factors include participation and inclusion of small-scale fish farmers and farmers associations in the planning and implementation of projects.

Constant interaction between the beneficiaries/target groups and projects is required (Abington, 1992).

There are two key elements that must be considered when assessing the suitability of fish farming. These include the viability of the fish production system and satisfaction of basic needs (both economic and social). According to Garling (2009) economic considerations can be classified into investment, finance, production and marketing. Net present value (NPV) is one of the popular methods for analyzing profitability of an investment project. This gives an indication of the present value of future earnings. If the NPV of a project is positive, the project is profitable and if the project is negative it should be abandoned (Sobo, 2014). The higher the NPV value the more profitable the business/investment is.

There are three main economical areas that fish farming can be of help. These areas include:

- Growth.
- Trade.
- Standards of living.

The final product quality and the individual task times are the major components of sustainability.

8.3.4. The Cost of Inputs and Sustainability of Small Scale Fish Farming

According to studies, there are numerous challenges that face fish farming activities in the world. For any fish-farming project to be sustainable and productive, there is the need to apply modern technological methods (Ofosu, 2011). There is also the need for proper management of all the inputs in this activity. There are several inputs in a fish farming business. Some of the most common inputs include:

- Breeding stock.
- Pesticide.
- Feed.
- Vaccine.

One of the biggest constraints to fish farming, especially in developing countries, has always been lack of quality breeding stock. Breeding stock usually plays a crucial role in fish farming and it is important for fish farmers to use good quality breeding stocks. In other words, good quality breeding stocks will guarantee the following:

- Fast growth.
- High yields.
- Good survival rate.

Due to high demand of fingerlings for fish farming, there are numerous privately owned hatcheries which produce and distribute these fingerlings. Therefore, it is important as a fish farmer to ensure that you get your breeding stock from reputable hatcher who is authenticated by government. This will ensure quality fingerlings production (Acton, 2013).

There are several technologies that have been adopted in order to increase fish production in fish farming by promoting fish seed production. Some of these technologies include genetic improvement and hormonal manipulation. These technologies have helped a lot in production of appropriate fingerlings that are resilience to various environmental conditions.

Research shows that the success of commercial fish farming operation is of paramount importance to all fish farmers. For fish farmers to achieve this success therefore they must have access to well balanced and cost-effective feeds coupled with optimal on-farm feed management practice.

Most fish farmers depend heavily on natural foods in the pond system with some supplement feed to increase fish yield (Tidwell, James, & Allan, 2001). This is because most fish feeds are imported and usually sold at very high prices due to high taxation and VAT. Some are contaminated and their quality tampered with. Some farmers have resulted to using supplement feeds that are made of locally available raw materials such as cassava leaves, sunflower, and white bran among others to feed their fish.

There are other feeding strategies that can be used in fish farming. These include:

- Spreading feed at fixed points at the same time daily.
- Bag and restrict feeding techniques.
- Promoting natural pond productivity.
- Break feeding schedule.

8.3.5. Extension Services and Sustainability of Small Scale Fish Farming

Provision of extension services to rural farmers as a way of enhancing their fish farming techniques is very significant. Other than enhancing their skills, the following are other benefits of extension services in fish farming:

- Providing fish farmers with information on production.
- Helps in value addition.
- Provide information on access to finance.
- Helps in educating them on profitable marketing of their fish produces.

Therefore, provision of extension services plays an important role in sustainability of fish farming in the society. According to FAO (2004) report, fish farming extension services have played an important role in the development of this sector nevertheless more is expected in the future. At grass root level, efficient extension services are required to promote the existing farmers and potential farmers for effective promotion of equitable and sustainable development of aquaculture sector.

Studies show that the effectiveness of extension services has declined over the last few years mainly because of a sharp reduction in operational budgets and human resource in the government. According to FAO (2004), one of the main challenges in extension services is low level of support to field technician, especially those staffs who are in direct contact with fish farmers on the ground.

Another major challenge facing extension services delivery in fish farming sector is inadequate extension officers provided by the department of fisheries. Governments also need to offer support services tailored towards enabling small-scale fish farmers take up commercial farming through sustainable practices.

Other recommendations on how to enhance fish farming include:

- Restructuring extension services to make technology dissemination responsive to small-scale fish farmers.
- Capacity building and skills training in order to enhance the success of fish farming development.
- Providing fish farming guidelines to small-scale fish farmers.

Studies have shown that engaging in sustainable fish farming activities helps in reducing poverty. Therefore, extension services should be regarded as the key enhancement of competency of fish farmers and local communities.

FAO (2004) report revealed that "quality extension services" does not simply mean that technicians have adequate technical training. In most of the African countries for instance, fish farming systems are characterized by high-cost of inputs, inadequate extension services and lack of markets all leading to low fish production in the region. Despite the fact that most governments are aware of the insufficient extension services, a lot of government and donor-funded projects are being implemented before addressing the issue of inadequate and trained extension workers.

8.3.6. Market Accessibility and Sustainability of Small Scale Fish Farming

In most developing countries, development of fish farming has been occurring in recent years and marketing of fish products has been an area of concern with many farmers desiring to attract international and regional markets.

In Africa for instance, the domestic market is too poorly organized to take advantage of the regional market. According to FAO (2016) report, fish market in Africa is classified according to target markets-local and international market-where local fish is largely sold fresh while external market involves high-quality standards during handling, processing and storage. Some of the main local fish markets where most of the small-scale farmers sell their produce include supermarkets, hotels, schools, fish outlet center. Some African countries like Kenya are also known to export farmed fish to the EU markets. This has opened up more market opportunities of fish and fisheries products to the European countries.

Lack of market for fish produce in Africa and other developing regions is one of the main reasons why most fish farmers cannot improve their living standards (OECD, 2003). There are many challenges linked to poor market accessibility. Some of these challenges include:

- Price risk and uncertainty.
- Lack of organized small-scale fish producers which increases the cost of putting together sparsely dispersed quantities of produce.
- Lack of meeting the required market standards.
- Poor roads (this leads to poor domestic markets with little or no

room for special or temporal integration, low prices and weak international competitiveness).

- Farmers not able to produce enough to meet the market demand.

Most of these challenges can be solved through Fish farmers embracing cluster production and marketing approaches through cooperative societies to benefit from economic of scale. It is also important that fish production to be linked to market demand and fish farmers must ensure that they produce products demanded by the market.

In South Africa for instance, fish farmers have formed cooperatives and collective marketing associations in order to increase their bargaining power in the market.

8.3.7. New Technology and Sustainability of Small Scale Fish Farming

Embracing and using technology for sustainable fish farming systems is a multi-disciplinary approach taking into account a wide range of objective geared towards sustainable fish farming. Globally, technologies are increasingly being developed for farm level application with an impact on the sustainability beyond the farm.

According to OECD/FAO (2017), over the last five years the system and technology used in fish farming has developed rapidly. Recent advances in information technology have also had profound influence on all walks of life and fish farming sector is no exception. Therefore, the growing significance of fish farming as an alternative source for food has further emphasized the need to adapt and develop advanced technologies for the better management of fish farming facilities as well as the regional planning for development. improving farm production through integrated modern technologies into the existing farming systems is essential for the enhancement of household food and income security.

It has therefore been suggested that fish farmers should strive to improve the profitability of fish farming through embracing new technologies. Some of the benefits of technologies in fish farming include:

- Reducing the risk of losing fish.
- Shortening culture cycle to target market size fish.
- Helps to lower cost of inputs.
- Helps in integrating fish farming with the existing farming systems.

- Helps in urban market accessibility.

However, there are several factors that limit the use of technology in fish farming by most farmers. Some of these factors include:

- Lack of awareness.
- Lack of knowledge of effects of recommended technology.
- Negative attitude to the innovation.

The use and uptake of new technologies in fish farming can therefore be enhanced through extension services to ensure sustainability of small-scale fish farming projects.

It has been noted that one of the reasons for slow fish farming development in developing countries has been; use of traditional fish and water husbandry, political, social and economic constraints that restrict investment and delay expansion and lack of information on fish farming technology. Studies have shown that adoption of fish farming technology is more likely to be adopted by the younger farmers. However, it is difficult for some countries to obtain knowledge on pond design and construction, hatchery equipment and other farm inputs such as aerators, cages and hatching incubators (Tidwell, James, & Allan, 2001).

In regards to the use of technology, it has been noted that feed management, selection and management of breeding stock are some of the important technological components in fish farming. In developing countries, most fish farmers still use traditional methods of fish farming, harvesting and preservation. Lack of technology has led to reduced output as well losses since the fish harvested cannot be stored for a long period that would enable fish farmers to market their produce at a later date or even transport to a different location for sale (Allison, Edward, & Ellis, 2001).

Fish farming has very high potential which can be fully used if only technology was adopted. Research has shown that fish farmers adopt fish farming technologies if they are assured that fish farming is a profitable venture. Therefore, the importance of the recommended fish farming technology related to existing practices must clearly be demonstrated to farmers. This will ensure that the small-scale farmers get the desired benefit, low-cost technologies appropriate to the farmers needs to be extended widely. Constant information on new innovations made by research institutions can also help in drawing benefits from the innovations (Allison, Edward, & Ellis, 2001).

9 CHAPTER

ARID AND SEMI-ARID AREAS POTENTIALS IN AGRICULTURAL SUSTAINABILITY

CONTENTS

9.1. Introduction 214

9.2. Major Agricultural Potential In Arid And Semi-Arid Areas 215

9.3. Key Constraints Tto Agricultural Production in Arid and Semi-Arid Areas 217

9.4. Key Strategies in Tapping Arid and Semi-Arid Agricultural Potentials 219

9.5. Food Security in Arid and Semi-Arid Areas 225

9.6. Solutions Towards Agricultural Food Sustainability in Arid and Semi-Arid Areas 225

9.7. The Factors Influencing Success and Failure in Agriculture 227

9.1. INTRODUCTION

This chapter is going to analyze the potentiality and sustainability of agricultural practices in arid and semi-arid areas. It has emerged that the ration of land to population is adversely low in most arid and semi-arid areas. Most of the land in these areas can seldom support populations. Therefore, it is very critical to seek solutions to the low productivity of these areas, to minimize the threat of famine, and increase sustainability in terms of agricultural production. Some of the main topics discussed in this chapter include: major agricultural potential in arid and semi-arid areas; key constraints to agricultural production in arid and semi-arid areas; key strategies in tapping arid and semi-arid agricultural potentials; food security in arid and semi-arid areas; solutions towards agricultural food sustainability in arid and semi-arid areas; and factors influencing success and failure in agriculture (Table 9.1).

Table 9.1: Agricultural Potential Concept in Arid and Semi-Arid Areas

Agricultural production in arid and semi-arid areas		
Types of agricultural productions	**Agricultural crops production** • Rain-fed production • Irrigated production	**Agricultural livestock production** • Animal rearing • Beekeeping
Constraints in agricultural productions	• Socio-economic constraints • Physical/environmental constraints • Technological constraints	
Strategies	• Irrigation • Credit provision • Appropriate technology • Provision of extension services • Formation of corporates • Community-based organizations • Introducing water catchment techniques	
Possible solutions	• Farm tools and equipment • Introduction of tolerant crops • Introduction of tolerant animals • Appropriate technologies • Accessibility to rural areas • Timely farming activities • Cooperatives formation • Accessibility to affordable credit services • Introduction to irrigation schemes • Land carrying capacity maintenance	
Results/outputs	Integrated agriculture for sustainable development	

Source: http://erepository.uonbi.ac.ke/bitstream/handle/11295/28562/.

9.2. MAJOR AGRICULTURAL POTENTIAL IN ARID AND SEMI-ARID AREAS

Globally, most arid and semi-arid areas are normally considered as potential albeit. Livestock keeping is one of the economic potentials in these areas. Statistics show that livestock industry in Africa's arid and semi-arid areas contributes millions of dollars per year in terms of foreign currency. This money comes through export of meat, milk, hides and breeding stock (FAO, 1993) (Figure 9.1).

Figure 9.1: Livestock keeping in arid and semi-arid areas (*Source:* https://www.treehugger.com/green-investments/sustainable-cattle-kenya-pays.html).

Most arid and semi-arid areas are normally not suitable for agricultural food production, especially through rain-fed production. Apart from erratic and unreliable rainfall, other factors that hinder crops food production in these areas include:

- Inadequate water supply.
- High-cost of land preparation for agricultural activities.
- Competition of the livestock activities for land resources.
- Limitations of the energy sources and implements used.
- Inadequate adoption of the available biochemical inputs (such as improved seeds and chemical fertilizers).

- Resistance to embrace certain draft resistant crops arising from dietary preferences.

However, recent studies show that there is an extremely high population growth rate in some arid and semi-arid areas. This is mainly attributed to: migration from high potential zones undergoing population pressure; internal movement within arid and Semi-Arid areas due to the carrying capacities being reached or exceeded; establishment of settlement or irrigation schemes; and subdivision of large-scale farms (Tejero & Zuazo, 2018) (Figure 9.2).

Figure 9.2: Vegetable production in arid and semi-arid areas (*Source:* http://andrewnyamuconservation.blogspot.co.ke/2016/08/water-efficient-vegetable-farming-for.html).

The demand for cultivation and grazing land contributes to illegal settlement (Squatting), one type of which is found on special reserves of state land and on the margins of game parks, on land owned by pastoral tribes and on large-scale ranches and farms in the arid and semi-arid areas . Migrants often apply technologies inappropriate to their newly acquired land, leading to further deterioration of the fragile ecosystems. Other effects of migrants into these areas include:

- Land devastation.
- Declining crop yields.
- Deteriorating farm systems diminish the quality and quantity of food produced.

Because of the hostile environmental conditions found in arid and semi-arid areas, inhabitants must adapt to those dynamic and fluid environments which incorporates complex values, beliefs and adjustment mechanisms.

Dutta (2011) noted that most of them view their economic functions as complementary to social ones since survival in such a harsh environment depends at least partially, on traditional social insurance systems. Most arid and semi-arid areas can be classified into different zones. These include:

Zone IV: These are arid and semi-arid areas with a moisture index of 30–42. They normally receive an average of between 500–800 millimeters of rainfall in a year.

Zone V: These are arid areas whose moisture index ranges from 42–51. They usually receive an average of between 300–500 millimeters of rainfall in a year.

Zone VI: This zone is normally classified as very arid. Its moisture index normally ranges from 51 – 57, while average rainfall per year ranges from around 200–350 millimeters.

9.3. KEY CONSTRAINTS TO AGRICULTURAL PRODUCTION IN ARID AND SEMI-ARID AREAS

Most of the constraints to development of arid and semi-arid are can be classified as socio-economic, physical/environmental and technological.

9.3.1. Social Constraint

These constraints normally arise from the cultural aspects of the people which affect the tapping of arid and semi-arid land resources negatively.

9.3.2. Economic Constraints

Most of the economic constraints normally arise from low incomes with the majority of the arid and semi-arid land's population. Other economic constraints are lack of security of land tenure among others.

9.3.3. Physical/Environmental Constraints

There are several causes of physical/environmental constraints in arid and semi-arid areas. Some of these causes include:

- Low, unpredictable and erratically distributed rainfall
- Poor soil fertility
- Pests and diseases

- Under physical
- Rough terrain among others

9.3.4. Technological Constraints

Most of these arise from use of unsuitable technology by immigrants to exploit arid lands.

To effectively transform arid and semi-arid areas into productive lands, there are several limiting factors that must be dealt with first. First, there is the need to outline all the major constraints that must be overcome. Most of the environmental obstacles are the best known and the overriding constraints. Limited and erratic rainfall makes the present crop and animal production system marginal and risky in these areas at the present level of technology. The result is inadequate production to feed the population of the areas, creating a dependence on outside relief (Ruben, Pender, & Kuyvenhoven, 2007).

It has also been noted that scarce resources often result in fight between groups or individuals for access to these resources. Considering the limited land potential and water availability, these must be allocated in ways that encourage multiple uses yet do not seriously reduce the productive capacity of any one activity.

Belshaw (1979) showed that inadequate technology is a serious problem. Governmental programs have largely focused on the high potential areas. The plant varieties, farming systems, water and conservation technologies needed for full production use of these areas have not been fully developed or adopted in most countries, especially in Sub-Saharan Africa. This is one of the main limiting factors on agricultural production. Knowledge of arid and semi-arid areas and of the nature of the response to development in pastoral economies also limits development.

Another constraint that must be dealt with is population and migration constraints. These constraints are serious because present land and water endowments and their distribution cannot support increased population using present technologies. The problem is compounded by immigrants from higher potential areas who bring farm systems suitable for the areas they have left but which are potentially destructive to the environment in the semi-arid areas (Tejero & Zuazo, 2018).

9.4. KEY STRATEGIES IN TAPPING ARID AND SEMI-ARID AGRICULTURAL POTENTIALS

There are numerous strategies that can be put into practice in order to transform most of these areas in order to become productive. Some of these strategies are discussed in the following subsections.

9.4.1. Development of Dry Land Farming Systems

Research has shown that developmental programs are crucial in transforming these areas and become productive in one way or another. Most governments have initiated programs that involve farmers co-operating in on-farm trials and demonstrations and receiving free inputs and where justified, food aid in compensation for land and labor opportunity costs.

Some of the viable or proven components of farming systems that can be adopted in arid and semi-arid areas include

- Agro-forestry species new drought-resistant maize and sorghum varieties, the latter grown on rationing mode for seed and drought period food reserve, grain legumes, root crops and oil seeds
- It is also important to encourage vertically integrated farming systems. This is because types of farming systems help in providing accelerated access to improved farm incomes.
- Another area that requires to be expanded is on oxen training and supply of ox-drawn equipment including ox-carts for farm-market access.
- Soil and water conservation methods such as terracing and Wocat water conservation technologies should also be integrated within the improved farming systems (Figure 9.3).

Terracing technology Wocat technology

Figure 9.3: Two key water conservation technologies used in arid and semi-arid areas (*Source:* https://qcat.wocat.net/en/wocat/technologies/view/technologies_487/ & https://commons.wikimedia.org/wiki/File:Pisaq_02.JPG).

Other methods for conserving soil and water in arid and semi-arid areas that need to be enhanced include:

- Vegetative cover in inter-cropping.
- Relay cropping.
- Organic structure systems.
- Improvement for water retention such as composting.
- The use of green manure.
- Mulching.

The government should also ensure that the role played by livestock keeping, especially when it comes to food security systems is strengthened and enhanced. There are several ways in which livestock farming in arid and semi-arid areas can be strengthened and enhanced. Some of these ways include:

- Practicing reliable food cropping as an alternative to livestock farming.
- Value-addition from livestock products.
- Reliable famine relief and post-drought recovery programs.
- Incorporating breeding stock entitlement and food-for-work components.

9.4.2. Beekeeping

Several governments have proposed beekeeping as one of the strategies important in tapping the potentiality of arid and semi-arid areas. Gupta et al. (2014) have shown that beekeeping can be used as a method of increasing incomes for the residents of arid and semi-arid areas, especially those that do not have adequate land for crops or livestock farming. Research done in arid and semi-arid areas have shown that modern beekeeping is mostly suited to women groups and lessons from traditional beekeepers are being carefully assimilated (Figure 9.4).

Figure 9.4: Beekeeping in arid and semi-arid areas (*Source:* http://desertswarm-bees.com/beekeeping/).

9.4.3. Small-Scale Irrigation schemes

Globally, irrigation schemes have been used as the best alternative methods of enhancing agricultural crops farming in rain-fed areas. Since most of the arid and semi-arid areas normally receive very minimal rainfall that cannot support agricultural crop production, irrigation schemes can be adopted as alternative methods of farming (Figure 9.5).

Figure 9.5: Irrigation schemes in arid and semi-arid areas (*Source:* http://news.mit.edu/2015/quantifying-impacts-of-large-scale-irrigation-on-rainfall-0907).

The best approach is to adopt accessible irrigation technology to the needs of smaller units. New irrigation schemes, rehabilitation of older ones and improved maintenance of irrigation works should be accorded high priority in these areas (Cantor, 1967).

More efficient use of water for agricultural production in arid and semi-arid areas can be safeguarded by water harvesting techniques of all kinds in areas far from river basins.

Those areas that receive considerably higher rainfall should adopt the development of various water conservation structures especially the integrated network of drains and small dams constructed in order to provide high pay-off supplementary irrigation water. On the other hand, those areas that receive lower rainfall, new techniques such as triangular and trapezoidal bands can be applied where appropriate.

Studies have shown that ground-water resources have so far not been used to any significant extent in the arid and semi-arid areas. Therefore, the government should invest into technologies that will help in the economic delivery of water to plants which make effective use of the restricted supply of ground-water in the form of drip irrigation (Dutta, 2011).

9.4.4. Proper Planning and Management of Arid and Semi-Arid Areas

The Ministry of Planning and National Development plays a significant role of planning and development in a country. Arid and semi-arid areas require proper planning and development in order to become productive (Lichfield, 2001). One country that has really capitalized in planning and development of arid and semi-arid areas is Israel (specifically in Negev which is one of the largest desert areas in Israel) (Figure 9.6).

Figure 9.6: Illustrating some of the achievement Israel has achieved in terms of revolutionizing desert land (*Source:* https://www.redduckpost.com/asian-farmers-are-learning-the-secrets-to-success-in-israel/).

Through the Jewish National Fund, the Israeli Government has adopted a variety of strategies to fight the desert conditions in Negev. Some of these strategies include:

Initiating afforestation programs: These programs have played a key role in semi-arid zones in Negev. Long term research and development in desert ecology in Israel has shown that by proper management, the process of desertification cannot only be halted, but can be reserved. Planting trees in desert lands has some major aims and it is same for all the countries. These include:

- Providing some greenery and shade near settlements.
- Establishing shelter belts around settlements.
- Protecting the soil and roads against water and wind.
- Providing fuelwood, charcoal and other minor products.

Geographically, Israel is a small country with high population. The government came into conclusion that the only available land in which to expand Israeli agricultural base and disperse the population, thereby easing the pressure on the central part of the country lies in the desert. Therefore, the government gave Negev high priorities in making it attractive and more productive.

Therefore, creation of a pleasing visual contrast to the barren dry landscape as well as the creation of a healthy ecological system is of utmost importance. Planting trees in arid and semi-arid areas are normally based on the principle of catching runoff. According to Habitat (1990) report, capturing the runoff water may account for up to 60% of the total precipitation also reduces soil erosion and preserves agricultural lands.

Developing water sources: This strategy can be viewed as an essential part of land development (Habitat 1990). It is important to note that the more water is available, the more land can be developed for agricultural use and the creation of pleasant environments. The existing conventional water sources in Israel are few, so nonconventional sources are being developed in a number of ways such as: construction of dams for catchment of runoff water; and constructing dams for enhancement of the existing aquifer by stopping the flood water flow and letting it seep into the ground for subsequent pumping.

Treatment and recycling wastewaters: Most effective use of recycled wastewaters in Israel that cannot be used in agricultural use is used in groves and parks. About 200 hectares of parks and groves are currently being irrigated with recycled wastewaters (Habitat, 1990). Brackish water, saline water can be used. The soluble salts are being used for irrigation purposes, either for salt tolerant agricultural crops or for planting trees and establishing recreational spots mainly in the Negev Highlands.

Savannization: This is a process that involves creation of a relatively high productive eco-system within the productive desert to raise the value of otherwise barren land. Normally, savannization helps to replace shrubs and soil crust associated with the desert with trees and grasses. This process is practiced in Negev and it can only be developed by implementing water harvesting techniques in which water run-off is used to supplement scarce rainwater. Some of the main techniques used in savannization of arid and semi-arid areas include:

- Agro-technical activities for runoff harvesting.
- Soil conservation activities.
- Tree planting.
- Increasing the productivity of plants and animals.

9.5. FOOD SECURITY IN ARID AND SEMI-ARID AREAS

According to World Bank (2003), the widespread problem of food insecurity in many developing countries are threatening the global economic, political and social stability. Sub-Saharan Africa is more affected by food insecurity than any other region in the developing world (FAO, 2004). The largest group of the hungry comprises members of households with low and variable income, limited assets, and few marketable skills (FAO, 1999). Most of the households are peasant farmers with limited land resources, unskilled and unemployed.

Nevertheless, empirical studies show that human factors on food security have comparatively been underemphasized compared with physical factors such as rainfall characteristics and soil productivity (FAO, 1997). Although physical environment places limits to resource exploitation, humans are capable of overcoming such limits through knowledge, capital and relevant technology. Subsequently, addressing human factors and institutional intervention is likely to shed more light towards improvement in food security.

According to FAO (1999) report on household food production in Sub-Saharan Africa, most social factors and cultural practices have a substantial impact on crop choice and yields. This indicates that certain food crops are neglected because of their taste or negative attitudes towards them. Subsequently, they end up disappearing from farmer's crop combination, thereby leading to chances of food insecurity. This situation has been observed in several places where food crops such as millet, sorghum and cassava have drastically declined in yields despite the fact that they are ideal in marginal areas.

9.6. SOLUTIONS TOWARDS AGRICULTURAL FOOD SUSTAINABILITY IN ARID AND SEMI-ARID AREAS

Farming technology has been examined as an important economic component in improving farm production and consequently food security. Semi-arid areas with food deficit could be transformed into self-sustaining net food producer by replacing in part or in whole, the traditional farming methods suitable to the area. Some of the techniques that can be used to enhance agricultural productivity and sustainability in arid and semi-arid areas include:

- Early land preparation.
- Early planting.
- Timely weed control.

However, for various reasons, many farmers are unable to modernize their farm enterprises and subsequently this affect food security. One of the common technologies that can be used in arid and semi-arid areas to enhance crop production is the oxen-driven plow. Oxen plow system is labor saving per hectare as compared to hoc cultivation systems. However, little information exists on the use and ownership of this technology among household in semi-arid areas. This technology is important, especially when it comes to timely planting and weeding in semi-arid areas where delay of one week can cause significant difference in crop yields. Therefore, farmers in arid and semi-arid areas should be encouraged and trained on how to use technologies such as oxen plow, which is appropriate in semi-arid area (Esteve, Brebbia, & Rico, 2008) (Figure 9.7).

Figure 9.7: A farmer using an oxen plow system to prepare his land for crops production (*Source:* http://news.hopethiopia.com/2012/05/09/touring-agricultural-projects-in-ethiopia/).

Access to financial facilities is another very crucial element in enhancing agricultural productivity and sustainability in arid and semi-arid areas. Lack of access to capital and credit is a major limitation to most farmers' contribution towards solving problem of food insecurity in arid and semi-arid areas (World Bank, 1999).

The government has a crucial role to play in formulation and implementation of policies that are likely to improve status of crop production in arid and semi-arid areas. The international organizations, NGOs, and other private institutions interventions in agricultural food production in these areas are also vital. Some of the critical roles played by these institutions include:

- Sharing of information of the agricultural food situation.
- Mobilizing essential resources (such as food, water, seeds and other donations) to mitigate the effects of famine.
- Capacity building of communities in fanning technologies.
- Provision of extension advisory services.

However, the involvement of these institutions typically lack coordination and timing in service delivery hence may lead to duplication of the same activities in one place or bias in the manner services are delivered.

Good agricultural food policy should ensure that all farmers are able to access improved seeds, agrochemicals, fertilizers and farm implements on credit or subsidized terms. In addition, it also must involve facilitating extension services to help farmers on how to improve farm production. Such policies are crucial in semi-arid areas where environmental conditions are harsh and household incomes are low (OECD, 2003).

Governments have a responsibility of developing mechanisms of directing agricultural financing and farming knowledge to all those farmers that are disadvantaged, for instance, those in areas like arid and semi-arid.

9.7. THE FACTORS INFLUENCING SUCCESS AND FAILURE IN AGRICULTURE

There are several factors that may contribute to agrarian development in arid and semi-arid areas. Among these are:

- Land reform.
- Introduction of high-value crops.
- Use of farm inputs.

Most farmers do not have adequate supply essential agricultural tools and equipment such as shovels, rakes, traditional hoes, wheel-barrows and ox-plows. Recent studies in Africa have revealed that most African farmers lack an appropriate supply of plows and as a result many farmers are unable to plow their land on time, and this prevents them planting their agricultural crops at the optimum time just as the rains are starting (Skinner, 1993).

Many farmers have to wait until a plow is available before they could plow and plant their crops.

According to research, it has been shown that a delay of 10 days in planting can lead to a 52% reduction in yield. Furthermore, the use of draft animal power can reduce the time required to cultivate a farm by over 60% compared to cultivation by hand and generally the more mechanized, the higher output per unit acre of land under agricultural activity. Therefore, in order to raise productivity in arid and semi-arid areas, farmers need to have access to basic tools and farm implements (Jose, 1989).

There are many other simple tools and equipment which can improve performance of farmers such as rotary planters, weeders, scarifiers and simple motor powered equipment. The utilization of these is still very low among small-scale farmers in arid and semi-arid areas. With some training and/or demonstration, existing tools could also be put to more and better use, for example, plows could be used much more widely for weeding if crops are planted in lines (Esteve, Brebbia, & Rico, 2008).

In summary, there are two main types of agricultural production potentials. These are agricultural crop production potential and livestock production potential. Under crop production, there is a potential in both rain-fed and irrigated agriculture while under livestock production, there is a potential in both animal rearing and bee-keeping. These potentials have not been fully tapped due to a number of non-agricultural constraints. These constraints are of socio-economic, physical/environmental, and technological in nature. Under socio-economic constraints there are such factors like small land sizes, lack of land tenure, and low incomes. Under environmental factors there are such factors like inadequate, erratic and unreliable rainfall, poor soils, rough terrain and high incidence of pests and diseases. Under technological constraints, there are such factors like lack of modern appropriate technology or use of inappropriate technology in agriculture.

However, to tap this agricultural potential in arid and semi-arid areas, a number of strategies may be pursued. These include: introduction of irrigated agriculture, accessing credit to the farmers, use of appropriate technology, provision of extension services to the farmers, formation of co-operatives and formation of Community-based organizations (CBOs). Possible solutions to fully tap the agricultural potential in arid and semi-arid areas may include among others: provision of crops to the farmers; growing and rising of drought, pest and disease resistant crops and animals respectively; timely planting, weeding and harvesting of the farmers crops; use of appropriate modern or traditional technology; introduction of irrigation agriculture; formation of farmers co-operatives and Community-based organizations; accessibility to agricultural development credit; and maintenance of land carrying capacity.

The output is to have an integrated agriculture in arid and semi-arid areas. This would result in employment, increased incomes, high standards of living and sustainable development. The various strategies are to be applied by various actors in different zones in arid and semi-arid areas.

10
CHAPTER

ORGANIC FARMING AND AGRICULTURAL SUSTAINABILITY

CONTENTS

10.1. Introduction 232
10.2. Sustainability of Organic Farming 232
10.3. The Start of Organic Farming 233
10.4. Elements of Organic Farming 235
10.5. Importance of Vertical Integration In Sustainable Organic Farming 236
10.6. Advantages of Organic Farming 237
10.7. Challenges In Organic Farming 238

10.1. INTRODUCTION

Organic farming is one of the many practices of sustainable agriculture. This chapter has very significant information about this practice that can be beneficial to all farmers. Some of the main subtopics discussed include: sustainability of organic farming; the start of organic farming; elements of organic farming; importance of vertical integration in sustainable organic farming; advantages of organic farming; and challenges in organic farming.

10.2. SUSTAINABILITY OF ORGANIC FARMING

According to Bellon & Penvern (2014), organic farming is considered as a sustainable solution to land loss among many poor farmers, especially developing countries. Most farmers in developed countries such as America and Europe have expanded to organic farming because of it numerous benefits. For instance, organic farming has received attention over the past decade because of environmental considerations and health issues of consumers; however, organic farming has not received extensive support from the agricultural sector because industry should revisit the existing economic models and revise them as necessary or, if required, develop a new model to meet the needs of farmers, especially socially disadvantaged farmers in developing countries (Bellon & Penvern, 2014) (Figure 10.1).

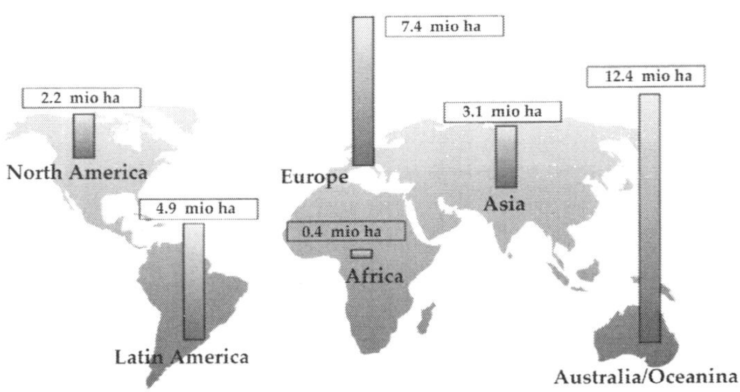

Figure 10.1: World scenario of organic farming (*Source:* https://www.slideshare.net/rajesh1128/rajeshorganic-farmin-presentation).

There are several issues faced by poor farmers both in developed and developing countries. Some of these issues hinder exploitation of organic farming in one way or another. These include:

- Lack of provisions and federal programs for socially disadvantaged farmers
- The majority of socially disadvantaged farmers are not aware of effective methods to sustain the farmers who are most at risk for land loss experience and their use of unsuccessful business models

To be successful organic farmers, there is the need for those socially disadvantaged farmers to be resourceful in their marketing techniques, such as using creative advertising. This will help them to reach specific audiences, especially when it comes to marketing of their organic agricultural products (Obach, 2015). There a number of strategies that can be used and they include:

- Internet marketing to reach consumers.
- Contacting schools, hospitals, nursing homes, and prisons.
- Establishing relationships with other farmers stands, restaurants, grocery stores, and processing and/or slaughtering plants for local.

According to Dahama (2002), changing economic trends and consumer demands have made organic farming a sustainable technique in the small-farming industry. It is important to note that most small-scale farmers can target the organic produce market better than large-scale producers can. This is because large-scale agricultural producers do not have the capacity to parcel out land to produce multiple types of crops. It has also been revealed that the maintenance of multiple varieties of produce requires changes in fertilizers, watering strategies, harvesting methods, and packaging, all of is a challenge to large-scale agricultural producers.

10.3. THE START OF ORGANIC FARMING

Since the beginning of the twenty-first century, most farmers especially in developed countries have shifted from conventional farming to organic farming practices. This has been contributed by consumer demands and changing economic markets which have brought organic farming into the forefront of the farming industry.

There have been numerous concerns regarding the use of chemicals in agricultural sector. Most farmers continue to use conventional methods of farming despite their economic disadvantages.

Gilkeson (2011) suggested that the general public need to be aware of the chemicals used in farming and how they can affect humans and the environment. Awareness of these issues is important especially policy changes and consumer demand for organic foods.

There are several benefits that come with organic farming practices in agricultural sector. Some of these benefits include:

- Soil erosion control.
- Increased productivity.
- Profitability.
- Improved soil quality.

In the 1980 s, a farm crisis occurred, with more farm foreclosures and bankruptcies occurring than at any other time since the Great Depression. Some conventional farms survived the crisis because consumers paid a premium for crops that were labeled organic because they had been neglected and therefore had not been treated with chemicals. These farms were some of the first to become organic farms. In the late 1980 s, the demand for and sale of organic produce increased. The overwhelming demand for organic produce resulted in Congress passing the Organic Food Productions Act in 1990 (Nandwani, 2016).

Leaders of the USDA created the National Organic Program (NOP) in 1992. The purpose of the NOP was to ensure that products labeled as organic always met organic requirements and comply with global policies for organic food. The NOP also includes regulations regarding livestock and crop management after harvesting, such as which materials are approved for use during processing and handling. Leaders of the NOP are tasked with the following functions regarding the organic certification process:

- Implementing policies, programs, and amendments within the NOP quality-management system.
- Managing personnel and the NOP budget.
- Organizing personnel and legislative affairs.
- Directing educational efforts for certifying agents and state programs, reinstating suspended certifications and accreditations, and managing the NOP internal team and program infrastructure.
- Creating and monitoring agreements with foreign governments; and

- Managing cost-share programs, supervising organic compliance and implementation, and examining alleged violations of NOP rules.

10.4. ELEMENTS OF ORGANIC FARMING

The basics of farming organically include:

- Using crop covers, excrement from animals, and crop rotation.
- Biological control techniques are used to manage weeds, fight insects, and decrease the risk of disease. There is no use of synthetic pesticides, fertilizers, hormones, and antibiotics in organic farming.
- Organic farming also included mixed-forage pastures for livestock, pasture rotation, and alternative health care for the well-being of animals. The focus of organic farming is on sustainable solutions, such as land and irrigation protection and procedures to achieve, sustain, and improve environmental stability.

Most farmers involved in organic farming normally emphasize the integration of the concepts of biodiversity and ecological balance. There is a common belief among scientists and engineers that the key to future agricultural sustainability is biodiversity (Halberg & Müller, 2013). One of the greatest challenges confronting humanity lies in understanding biodiversity, being educated in its use, caring for it, and conserving it.

Obach (2015) revealed that organic farming can enhance plant and animal development as well as increase growth in crops more than conventional farming can. Scientifically, it has been proven that organically farmed land normally yield larger areas of semi-natural border plant life in addition to greater vegetation in cropland and woodland and greater fertility in crops. Some of the major semi-natural habitats include woodland, field margins, and hedgerows. Woodland consists of simple borders on conventional farms, and fields comprise of farmland in which crops are grown (Obach, 2015).

It has been found that the conventional farmlands normally contain more wooded areas and the organic farmlands contain more forestland. It can therefore be concluded that conventional farmers could increase vegetation variety in farmland by applying ecological farming methods.

Advancements in biotechnological innovation in agricultural sector have led to improvements in hybrid seed design and production, making crops more pest resistant and able to grow larger and faster. Further, new herbicides are more effective in protecting crops than previous herbicides were. However, such advances are mainly available to big farming enterprises, which are usually better able to afford the technology (Esteve, Brebbia, & Rico, 2008) (Figure 10.2).

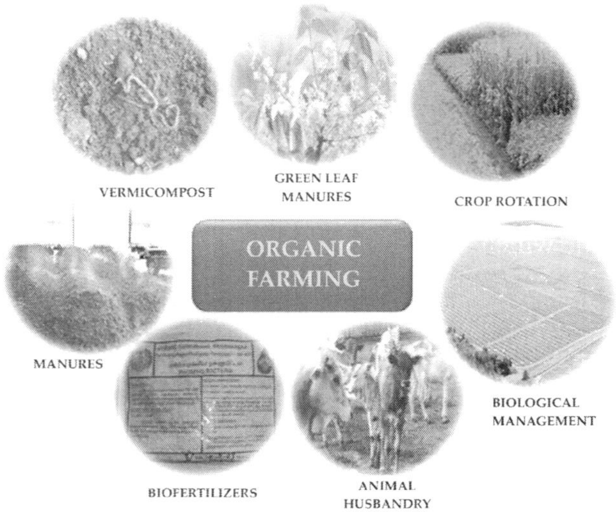

Figure 10.2: Key components of organic farming (*Source:* http://agritech.tnau.ac.in/org_farm/orgfarm_introduction.html).

10.5. IMPORTANCE OF VERTICAL INTEGRATION IN SUSTAINABLE ORGANIC FARMING

Vertical integration is a process in which a market system synchronizes by coordinating various phases of production and distribution. It requires a relationship among the farmer, buyer, and supply chain. There different reasons why farmers choose vertical integration. Some of these reasons include:

- Vertically integrated decision makers can quickly adapt to consumer demand.
- Through vertical integration, farmers have the resources to recover from natural catastrophes, and they have the financial capital to develop and invest in the latest biotechnology to improve productivity.

- Vertical integration is also supported by economic infrastructures nationally and locally.
- Communities benefit economically from the presence of large agribusinesses because of the increased employment opportunities, the expanded tax base, and the improved local infrastructures that such operations require.
- Vertical integration ensures that large agribusinesses have the first access to biotechnology and receive steep discounts.

Vertical integration moves control of farming manufacturing to food processing, and this system works well with managing risk. Because of the benefits resulting from vertical integration, the structure will likely continue to be supported, which will reinforce the barriers socially disadvantaged farmers face in the pursuit of sustainability. Farmers who use vertical integration will have an advantage over farmers who do not vertically integrate (Krishnamurthi, 2016).

10.6. ADVANTAGES OF ORGANIC FARMING

The demand for organic food had grown rapidly over the last 20 years. There are four major reasons why most farmers choose organic farming. These include:

- Organic farming can increase profitability because of government support, produce variation, and high consumer demand.
- Organic farming improves soil fertility, increases biodiversity, and raises environmental awareness.
- Farmers are concerned about the condition of farmland, nutrition, animal well-being,
- Organic farming can increase employment opportunities and economic growth in rural communities.

Obach (2015) revealed that organic farmers who grow wheat in the central and northern United States experience financial results that are equal to or better than those of conventional farmers. Because of organic farmers' diversified systems, the farmers are less likely to experience natural and economic risk than are convention farmers. Statistic shows that organic market has continued to grow in Europe and the United States. The highest percentage of growth has occurred in Denmark, Germany, Sweden, Italy, and Finland.

The reasons farmers in Europe have chosen organic farming include:

- the desire to produce chemical-free foods.
- to protect the fertility of soil.
- to provide employment opportunities.
- to provide healthy food to the local community.

Food safety is an important concern for organic farmers. Farmers believe they need to focus on alternative farming solutions because of the diseases that affected livestock in Great Britain in 1998.

It has been reported that the fertility of soil is determined based on the amount of organic matter in the soil. For instance, through organic farming, microbial activity in the soil increase by 30% and that biomass increase by 20% in European organic farms.

Personal health is a major concern among organic farmers and motivates conventional farmers to convert to organic farming. Obach (2015) showed that organic farming systems in the United States have proven to be sustainable farming methods. For instance, it has been reported that by using compost and manure instead of nonorganic materials will increase the profitability of organic farms.

Through crop diversification, farmers can help ensure the soil maintains its nutrients and thereby provide healthier produce to consumers. Most organic farmers firmly believe that synthetic products and pesticides are not good for the environment. They also believe that the health of the soil is improved by paying attention to the nutrient cycles of the soil. According to Krishnamurthi (2016), more than 70% of committed organic farmers choose organic farming for consumer health, own health, ethical reasons, and lifestyle preferences.

10.7. CHALLENGES IN ORGANIC FARMING

10.7.1. Legal Challenges

Most farmers choosing to convert to organic systems are normally constrained by legal requirements. To be certified organic, farmers must complete a conversion phase lasting at least 3 years. Many farmers abandon the process because of the financial strain and the long transition period (Bellon & Penvern, 2014).

10.7.2. Implementation Challenges

Farmers may also find organic farming methods difficult to implement. For instance, managing weeds without pesticides can be challenging. Some farmers also struggle to maintain buffer zones in organic areas.

10.7.3. Low Production

Numerous farmers experience lower yields when growing organic crops. Farmers might receive little or no guidance on organic farming and how to overcome the difficulties (Gilkeson, 2011). However, some farmers choose to convert despite understanding they may experience multiple challenges and disadvantages. Furthermore, other than experiencing reduced crop production and pest management challenges, transitioning farmers may experience a reduction in animal production and functioning and a lack of information on organic farming.

10.7.4. Transition Effect

An additional disadvantage of organic farming is the organic transition effect. This entails an immediate decrease in produce and livestock functioning because of insufficient measures to manage pests and diseases. A related challenge in converting to organic farming is the need for equipment designed specifically for organic farming. According to Krishnamurthi (2016), a major disadvantage in organic farming is the decrease in crop yields during the transition period. Crop yields may slowly increase because of crop rotation. Researchers have documented a decline in crop yields during the 1st year after transitioning from conventional farming and then a recovery over the three following years.

10.7.5. Skills and Knowledge Challenges

Organic farming requires skill in and knowledge of organic farming processes. Transitioning farmers in developing countries have faced pressure to increase their yields.

10.7.6. Livestock Issues

Another difficulty found in organic farming is hygienic issues in livestock and decreased livestock production. Organic and transitioning farmers have noted the difficulty of organic animal management. For instance, researchers have found that organic weaning methods for calves normally cost 10 times more than conventional methods.

10.7.7. Financial Issues

Studies have shown that most problems experienced during transition relate to lack of governmental and institutional support. Organic farmers need to learn how to market their products to overcome their challenges. Many transitioning farmers normally find that they need significant capital investment to keep their farms operating. These farmers also find it difficult to obtain funding for added farming expenses.

10.7.8. Other Challenges

Biodiversity benefits are dependent on location, climate, crop type, species, and management practices. It has been noted that organic livestock requires access to shade, shelter, fresh air, and direct sunlight.

REFERENCES

1. Abington, J. B. (1992). *Sustainable Livestock Production in the Mountain Agro-ecosystem of Nepal*, (Issue 105). London: Food & Agriculture Organization, 77-131.
2. Acquaye, & Smith. (1965). Effects of Ground Covers and Fertilizers on Establishment and Yield of Cocoa on Clear-Felled Land in Ghana. *Experimental Agriculture 1*(2), 131–139.
3. Acton, Q. A. (2013). *Issues in Agribusiness and Agricultural Economics:* (2013 Edition). New York: Scholarly Editions, 23-37.
4. Adams, J. E. (1966). Influence of Mulches on Runoff, Erosion, and Soil Moisture Depletion 1. *Soil Science Society of America Journal 30*(1), 110–114.
5. Agboola, & Fayemi. (1971). *Effect of interplanted legumes and fertilizer treatment on the major soil nutrients.* Int Symp Soil Fert Eval New Delhi Proc, 31-49.
6. Aland, & Madec. (2009). *Sustainable Animal Production: The Challenges and Potential Developments for Professional Farming.* Netherlands: Wageningen Academic Pub, 189-313.
7. Ali, H. (2011). *Practices of Irrigation & On-farm Water Management, Volume 2.* Berlin: Springer Science & Business Media.
8. Allison, Edward, & Ellis. (2001). The livelihoods approach and management of small-scale fisheries. *Marine Policy,* 377–388.

9. Anamosa, P. (1990). Water movement through an aggregated, gravelly oxisol from Cameroon. *Geoderma 46*(1–3), 263–281.
10. Aosa, E. (1992). *An Empirical Investigation of Aspects of Strategy Formulation and Implementaion Within Large, Private Manufacturing Companies in Kenya.* London: University of Strathclyde, 3-15.
11. Bellon, & Penvern. (2014). *Organic Farming, Prototype for Sustainable Agricultures.* Berlin: Springer Science & Business, 1-19.
12. Belshaw, D. (1979). Taking Indigenous *Technology Seriously: The Case of Inter□cropping Techniques in East Africa. IDS bulletin 10*(2), 24–27.
13. Benson, R. J. (1991). *Introduction to Mixed Farming.* London: Farm Management Home Study Program.
14. Bhullar. (2012). *Agricultural Sustainability: Progress and Prospects in Crop Research.* New York: Academic Press, 61-149.
15. Bluffstone, & Gunnar. (2012). *Agricultural Investment and Productivity: Building Sustainability in East Africa.* (1st Edition) London: Routledge, 52-91.
16. Campbell, S., & Moore, D. (1991). *The mulch book: a complete guide for gardeners.* Storey Communications, 68-94.
17. Cheeke, P. R. (1986). Potentials of rabbit production in tropical and subtropical agricultural systems. *Journal of Animal Science 63*(5), 1581–1586.
18. Christen, Squires, & Hudson. (2010). *Interdisciplinary and Sustainability Issues in Food and Agriculture, Vol 2.* London: EOLSS Publications, 12-34.
19. Dahama, A. K. (2002). *Organic Farming for Sustainable Agriculture.* New Delhi: Agrobios, 23-44.
20. Dillard, Dujon, & King. (2008). *Understanding the Social Dimension of Sustainability.* London: Routledge, 13-95.
21. Doneen, L. D. (1971). *Irrigation Practice and Water Management: 15136–71-M1.* Rome: UNO (FAO), 81-123.
22. Dris, & Jain. (2004). *Production Practices and Quality Assessment of Food Crops: Volume 1: Preharvest Practice.* Berlin: Springer Science & Business Media, 199-253.
23. Dutta, M. K. (2011). *Irrigation Potential in Agriculture of Assam.* New Delhi: Concept Publishing Company, 120-165.

24. Erickson, Frederick, & Gutierrez. (2002). *Comment: Culture, Rigor, and Science in Educational Research. Educational Researcher 31*(8), 21–24.
25. Esteve, Brebbia, & Rico. (2008). *Sustainable Irrigation Management, Technologies and Policies II.* London: WIT Press.
26. Potthast, T., & Meisch, S. (Eds.). (2012). *Climate Change and Sustainable Development: Ethical Perspectives on Land Use and Food Production: EurSAFE 2012, Tübingen, Germany, 30 May-2 June 2012.* Springer Science & Business Media, 120–157.
27. Evenson, & Rumbaugh. (1972). Influence of Mulch on Postharvest Soil Temperatures and Subsequent Regrowth of Alfalfa (Medicago sativa L.) 1. *Agronomy Journal 64*(2), 154–157.
28. Ewers, B. (1999). Carry-over effects of water and nutrient supply on water use of *Pinus taeda. Ecological Applications 9*(2), 513–525.
29. Faires, N. (2016). *The Ultimate Guide to Urban Farming: Sustainable Living in Your Home, Community, and Business.* New York: Skyhorse Publishing, 31-72
30. FAO. (1992). *Agricultural Sustainability: Definition and Implications for Agricultural and Trade Policy.* London: Food & Agriculture Org, Issue 106, 58-98.
31. FAO. (1993). *Sustainable Animal Production from Small Farm Systems in South-East Asia.* London: Food & Agriculture Org, 15-59.
32. FAO. (1997). *The State of Food and Agriculture 1997: (Vol 1).* New: Food & Agriculture Org, 441-489.
33. FAO. (2006). *Livestock Report 2006.* New York: Food & Agriculture Org, 3-29.
34. FAO. (2009). *The State of Food and Agriculture 2009: Livestock in the Balance.* New York: Food and Agriculture Organization of the United Nations, 49-61.
35. Fayeye, & Ayorinde. (2010). Effects of season, generation, number of mating, parity and doe number of teat on doe and litter birth characteristics in domestic rabbit. *Research Journal of Animal and Veterinary Sciences*, 6–9.
36. Fox, T. (2011). *Urban Farming: Sustainable City Living in Your Backyard, in Your Community, and in the World.* California: i5 Publishing, 370-408.

37. Gebremedhin, B. (2002). *Collective Action for Grazing Land Management in Mixed Crop-livestock Systems in the Highlands of Northern Ethiopia.* Addis Ababa: ILRI (aka ILCA and ILRAD), 3-19.
38. Ghosh, S. K. (2010). Effect of breed and season on rabbit production under subtropical climate. *World Rabbit Science,* 29–33.
39. Gibon, & Mihina. (2003). *Livestock Farming Systems in Central and Eastern Europe.* Netherlands: Wageningen Academic Pub, 25-73.
40. Gilkeson, L. (2011). *Backyard Bounty: The Complete Guide to Year-Round Organic Gardening in the Pacific Northwest.* New York: New Society Publishers, 29-63.
41. Grigg, D. B. (1974). *The Agricultural Systems of the World: An Evolutionary Approach.* Vol 1, London: Cambridge University Press, 112-152.
42. Gupta et al. (2014). *Beekeeping for Poverty Alleviation and Livelihood Security: Vol. 1: Technological Aspects of Beekeeping.* Berlin: Springer, 171-191.
43. Guyer, J. P. (2018). *An Introduction to Hydro-Mulching.* New York: Guyer Partners, 156-171.
44. Halberg, & Müller. (2013). *Organic Agriculture for Sustainable Livelihoods. 1st Edition*: London: Routledge, 21-51.
45. Hanks, Bowers, & Bark. (1961). Influence of soil surface conditions on net radiation, soil temperature, and evaporation. *Soil Science 91*(4), 233–238.
46. Hayami, & Ruttan. (1971). *Agricultural Development: an International Perspective.* London: Johns Hopkins Press, 212-256.
47. Henkel, M. (2015). *21st Century Homestead: Sustainable Agricultural.* New York: Lulu.com, 27-65.
48. Hester, & Harrison. (2005). *Sustainability in Agriculture.* London: Royal Society of Chemistry, 49-86.
49. IIED. (2001). *Measuring the Effects of Environmental Regulation in Oligopolistic Markets with Differentiated Products – 9035iied.* New York: IIED, 9-21.
50. Johnson, & Scholes. (2002). *Exploring Corporate Strategy with Pin Card.* London: Pearson Education, 32-59.
51. Kang, M. S. (2010). *Water and Agricultural Sustainability Strategies.* Florida: CRC Press, 3-43.

52. Kebreab, E. (2013). *Sustainable Animal Agriculture.* London: CABI, 124-172.
53. King'oriah, G. (1999.). *Proposals for Regional Cooperation in Africa.* Nairobi: No. INIS-XQ-003.
54. Koont, S. (2017). *Sustainable Urban Agriculture in Cuba.* Florida: University Press of Florida, pp.250.
55. Krishnamurthi, P. D. (2016). *Organic Agriculture for Sustainability.* Chennai: Notion Press, 31-44.
56. Lattanzi, & Baumgardner. (1974). Influences of Mulch Rate and Slope Steepness on Interrill Erosion 1. *Soil Science Society of America Journal 38*(6), 946–950.
57. Leitch, Isabella, & Godden. (n.d.) (1941). The efficiency of farm animals in the conversion of feeding stuffs to food for man. *The Efficiency of Farm Animals in the Conversion of Feeding Stuffs to Food for Man.* London: Imp. Bur. Animal Nutrit, pp.64 pp.
58. Lichfield, N. (2001). The philosophy and role of community impact evaluation in the planning system. *Recent Developments in Evaluation,* 153–173.
59. Lichtfouse, E. (2014). *Sustainable Agriculture Reviews, Volume 15.* Berlin: Springer, 15-55.
60. Loomis, & Connor. (1992). *Crop Ecology: Productivity and Management in Agricultural Systems.* London: Cambridge University Press, 23-69.
61. Macnaghten, & Ripalda. (2015). *Governing Agricultural Sustainability: Global Lessons from GM Crops.* London: Routledge, 6-39.
62. Markham. (2010). *Mini Farming: Self Sufficiency on a 1/4 Acre.* New York: Skyhorse Publishing Inc, 37-99.
63. McCalla, & Duley. (1946). Effect of crop residues on soil temperature. *Journal of the American Society of Agronomy.* Vol 38, 75-89.
64. McNeill, Nesheim, & Brouwer. (2012). *Land Use Policies for Sustainable Development: Exploring Integrated Assessment Approaches.* London: Edward Elgar Publishing, 11-69.
65. Morris, M. L. (2007). *Fertilizer use in African Agriculture: Lessons Learned and Good Practice Guidelines.* London: World Bank Publications, 34-51.

66. Mougeot, L. J. (2006). *Growing Better Cities: Urban Agriculture for Sustainable Development.* Ottawa: IDRC, 49-58.
67. Mustafa. (1999). *For Hunger-proof Cities: Sustainable Urban Food Systems.* London: IDRC, 41-84.
68. Nair, K. P. (2016). *The Nutrient Buffer Power Concept For Sustainable Agriculture:* (20th Edition) New Delhi: Notion Press, 255-260.
69. Nandwani, D. (2016). *Organic Farming for Sustainable Agriculture:* (Vol 9) Berlin: Springer, 123-136.
70. National Research Council. (1993). *Population and Land Use in Developing Countries: Report of a Workshop.* New York: National Academies Press, 30-70.
71. National Research Council. (2010). *Toward Sustainable Agricultural Systems in the 21st Century.* New York: National Academies Press, 43-78.
72. Obach, B. K. (2015). *Organic Struggle: The Movement for Sustainable Agriculture in the United States.* Massachusetts: MIT Press, 81-127.
73. OECD. (2000). *Towards Sustainable Development Indicators to Measure Progress (Proceedings of the Rome Conference): Indicators to Measure Progress (Proceedings of the Rome Conference).* Paris: OECD Publishing, 45-69.
74. OECD. (2003). *Organic Agriculture Sustainability, Markets and Policies: Sustainability, Markets and Policies.* Paris: OECD Publishing, 19-31.
75. OECD. (2003). *Organic Agriculture Sustainability, Markets and Policies: Sustainability, Markets and Policies.* Paris: OECD Publishing, 49-58.
76. OECD. (2006). *Water and Agriculture Sustainability, Markets and Policies: Sustainability, Markets and Policies.* Paris: OECD Publishing, 9-22.
77. Ofosu, E. A. (2011). *Sustainable Irrigation Development in the White Volta sub-Basin: UNESCO-IHE PhD Thesis.* Florida: CRC Press, 133-167.
78. Omosa, M. (1998). *Population Growth, Land Use, and Food Self-sufficiency in Kenya: A Comparative Analysis of Small and Medium-large Land Holdings in Kisii and Nyamira Districts:* (Vol 2). Johannesburg: Union for African Population Studies, 29-40.

79. Osei, J. (2009). *The Challenge of Sustaining Emergent Democracies: Insights for Religious Intellectuals & Leaders of Civil Society.* New York: Xlibris Corporation, 134-164.

80. Ozier-Lafontaine, & Lesueur-Jannoyer. (2014). *Sustainable Agriculture Reviews 14: Agroecology and Global Change.* Berlin: Springer, 53-117.

81. Papendick, Lindstrom, & Cochran. (1973). Soil Mulch Effects on Seedbed Temperature and Water During Fallow in Eastern Washington 1. *Soil Science Society of America Journal 37*(2), 307–314.

82. Parker, & Larson. (1962). Nitrification as Affected by Temperature and Moisture Content of Mulched Soils 1. *Soil Science Society of America Journal,* 238–242.

83. Parkhurst, Carmen, & Mountney. (2012). *Poultry Meat and Egg Production.* Berlin: Springer Science & Business Media, 151-172.

84. Pearson, C. (2011). *Urban Agriculture: Diverse Activities and Benefits for City Society.* (1st Edition) London: Routledge, 23-66.

85. Penman, H. (1963). *Vegetation and Hydrology: (Volume 96 - Issue 5)* New York: Forgotten Books, 122-154.

86. Pereira, Feddes, & Gilley. (2013). *Sustainability of Irrigated Agriculture. Berlin: Springer Science & Business Media, 540-588.*

87. Pereira, Feddes, & Lesaffre. (2013). *Sustainability of Irrigated Agriculture.* Berlin: Springer Science & Business Media, 431-473.

88. Peter, K. (2008). *Underutilized and Underexploited Horticultural Crops:* (Vol 4). New Delhi: New India Publishing, 31-61.

89. Pittenger, D. (2014). *California Master Gardener Handbook,* (2nd Edition). California: UCANR Publications, 37-83.

90. Pradeepkumar, T. (2008). *Management of Horticultural Crops, Part 1.* New Delhi: New India Publishing, 29-58.

91. Ragasa, Catherine, & Mariam. (2016). *The Impact of Agricultural Extension Services in the Context of a Heavily Subsidized Input System: The Case of Malawi:* (Vol 105), Lagos: Intl. Food Policy Res. Inst, 25-47.

92. Rosan, & Pearsall. (2017). *Growing a Sustainable City?: The Question of Urban Agriculture.* Toronto: University of Toronto Press, 3-61.

93. Roy, M. L. (2009). *Mixed Farming Amongst Resource-poor Farm Families in West Bengal-A Gender Perspective.* New York: NDRI, 33-50.

94. Ruben, Pender, & Kuyvenhoven. (2007). *Sustainable Poverty Reduction in Less-favoured Areas.* London: CABI, 63-133.

95. Snellen, W. B. (1996). *Irrigation Scheme Operation and Maintenance.* New York: Food & Agriculture Organization, 15-41.

96. Starrs, & Goin. (2010). *Field Guide to California Agriculture.* California: University of California Press, 41-80.

97. Stewart, & Coclanis. (2011). *Environmental Change and Agricultural Sustainability in the Mekong Delta.* Berlin: Springer Science & Business Media, 100-205.

98. Sutherland, Wilson, & Zagata. (2014). *Transition Pathways Towards Sustainability in Agriculture: Case Studies from Europe.* London: CABI, 33-83.

99. Takeshima, & Edeh. (2013). *Typology of Farm Households and Irrigation Systems: Some Evidence from Nigeria.* Lagos: Intl. Food Policy Res. Inst, 1-8.

100. Tei, Nicola, & Benincasa. (2017). *Advances in Research on Fertilization Management of Vegetable Crops.* Berlin: Springer, 137–148.

101. Tejero, & Zuazo. (2018). *Water Scarcity and Sustainable Agriculture in Semiarid Environment: Tools, Strategies, and Challenges for Woody Crops.* Netherlands: Elsevier Science, 66-104.

102. The Worldwatch Institute. (2012). *State of the World 2012: Moving Toward Sustainable Prosperity.* New York: Island Press, 94-132.

103. Tidwell, James, & Allan. (2001). *Fish as Food: Aquaculture's Contribution: Ecological and Economic Impacts and Contributions of Fish Farming and Capture Fisheries. EMBO Reports,* 958–963.

104. Toensmeier, E. (2007). *Perennial Vegetables: From Artichokes to Zuiki Taro. A Gardener's Guide to Over 100 Delicious and Easy to Grow Edibles.* New York: Chelsea Green Publishing, 26-72.

105. Trigo, E. (1991). *Toward a Strategy for Sustainable Agricultural Development.* London: Bib. Orton IICA/CATIE, 8-17.

106. Trzyna, & Osborn. (1995). *A Sustainable World: Defining and Measuring Sustainable Development.* California: Published for IUCN – the World Conservation Union by the International Center for the Environment and Public Policy, California Institute of Public Affairs.

107. van Loon, & Hugar. (2005). *Agricultural Sustainability: Strategies for Assessment.* London: SAGE, 56-105.

108. Veenhuizen, & Danso. (2007). *Profitability and Sustainability of Urban and Periurban Agriculture.* London: Food & Agriculture Organization, 40-59.

109. Verma, N. M. (1993). *Irrigation in India: Themes on Development, Planning, Performance and Management.* Delhi: M.D. Publications Pvt. Ltd, 37-85.

110. Willey, & Osiru. (1972). Studies on mixtures of maize and beans (*Phaseolus vulgaris*) with particular reference to plant population. *The Journal of Agricultural Science,* 517–529.

111. World Bank. (1999). *The Irrigation Sector.* Delhi: World Bank Publications, 44-72.

112. Wright, J. (2012). *Sustainable Agriculture and Food Security in an Era of Oil Scarcity: Lessons from Cuba.* London: Earthscan, 130-177.

113. Zhen & Zoebisch. (2006). *Resource Use and Agricultural Sustainability: Risks and Consequences of Intensive Cropping in China.* Berlin: Kassel University Press GmbH, 24-94.

INDEX

A

aeration 61, 74
agricultural enterprise 26, 29
Agricultural farming 6
agricultural land 4, 8, 14
Agricultural marketing system 159
agricultural productivity 13, 14, 16, 17, 32, 33, 34, 39, 43
Agricultural structure 28
agricultural sustainability 26, 28, 31, 33, 34, 35, 38, 39, 40, 42, 43
agricultural technology 34
Agricultural venture 3
aluminum painted gravel 93
Animal protein 188, 194, 195, 203
aquaculture 127, 139
artificial insemination 164
Ayrshire 153, 154
azotobacter 99

B

biodiversity 235, 237
biomass 139
biotechnology 2, 17, 18, 19, 20, 21, 22
black painted gravel 93
Body development 199
Brackish water 224
Breeding stock 215, 220
Brown Swiss 151, 152
Butchering 171

C

Cage housing system 190
Calf mortality 170
Calving intervals 162
capital investment 127, 134
Carcass composition 199
Cell Wall Constituents (CWC) 156
Civilization 108, 110
Climate change 27, 28

climbing crops 80, 81
Cognitive development 168
commercial fertilizers 102, 103
Community-based organizations (CBOs) 228
Crop fertilization 104
cropping intensity 78
Cropping Pattern 26
Crop production 60, 61, 221, 226, 227, 228
crops fertilization 80
crops pruning 80
Cultivation 216, 226, 228
Cysticerous bovis 180

D

dairy breeds 150, 151, 152
Dairy farming 150, 157, 158
denitrification 103
dietary deficiency 199
disposable surpluses 42
Docile temperament 154
Drip irrigation 222

E

earthworm 99
East Coast Fever (E.C.F.) 179
Ecological system 224
economic crisis 109, 115
economic incentives 43
Environmental health 131
Environmental planning and management (EPM) 119
Environmental pollution 103, 104
environmental threats 110
evaporation 94, 95
Evapotranspiration (ET) 63
Extensive system 197

F

Farming technology 225
Feed conversion 189
feedstuff 156
fertilizer 5, 17, 88, 102, 103, 104
financial branch 15
fish farming 203, 204, 205, 206, 207, 208, 209, 210, 211, 212
Fish farming system 204, 205, 206
Flood irrigation 55, 56
flowing streams 76
Food efficiency 191
food insecurity 110, 113
Food safety 238
Forage consumption 202
Formal channels 158
Fragmentation 31

G

Genetically Modified Organic crops (GMO) 33
genetic variation 164
genotype 200
Geographical information systems (GIS) 121
germination 100
Gestation 199
Global population growth 7
greenhouse gas emissions 169
greenhouse gasses 169
Gross Domestic Product (GDP) 169
groundwater 136, 137
Guernsey 152, 153

H

harvest times 40
Hereditary tradition 32

heterogeneity 160, 163, 164
heterogeneous herd variance 163
Heterogeneous phenotypic variation 163
Holstein 150, 151
Holstein dairy breed 150
horizontal trellis 80, 81, 82
horticulture 139
housing system 189, 190, 192
Human nutrition 168, 169
human resources 2

I

Indigenous beef cattle 185
Induced Innovation Hypothesis 4, 5
infiltration 94, 95
Inorganic fertilizer 118
insecticidal sprays 86
Intensive system 197
internodes 85
Irrigating system 57
irrigation 48, 49, 50, 51, 52, 53, 54, 55, 56, 57, 58, 59, 60, 61, 62, 63, 64, 65, 66, 69, 70, 71, 72, 73, 74, 75, 76, 77, 78
Irrigation protection 235

J

Jersey 151

L

Labor productivity 4
Land Fragmentation 26
Land productivity 5
leguminous crops 86
livestock 108, 109, 110, 111, 112, 114, 120, 124, 127, 128, 132, 133, 138, 139

M

magnesium deficiency 97
manure 88
Meat production 168, 175, 178
methane 169
Milk production 150, 151, 152, 154, 157, 160, 162, 165
Millennium Development Goals (MDGs) 34
mixed cropping 80, 86, 87, 88, 89
monocropping systems 87
mono-gastric animals 194
monopolistic powers 20
mulching 80, 90, 91, 93, 94, 95, 96, 97, 100, 101
Mulching material 96
Municipal solid waste (MSW) 138

N

National Organic Program (NOP) 234
natural resources 2, 3, 10, 27
nematode 99
Net present value (NPV) 207
nitrous oxide 169
non-biodegradable wastes 138
nutrition and management 163
Nutritious food 113

O

Optimal nutritional value 157
Organic farming 232, 233, 234, 235, 236, 237, 238, 239
organic matter 96, 97, 98
organic mulches 91
oribaticles 98
Over-fertilization 103

P

Paper mulch 93
paradox 40
Pasture rotation 235
permeability 61, 74
pests 84, 85
Phonolite lava 99
plant nutrients 96
polyethylene 93, 98, 100, 101
population growth 31
Population growth rate 216
Potential evapotranspiration (PET) 63
poultry 188, 189, 190, 191, 192, 193, 202
Production system 112
profitability 19

R

Rabbit meat production 195, 196, 199, 201
Rabbit production system 197
receptors 109
Regression techniques 160
Rehabilitation 52
reservoirs 53, 54, 76, 77
resource degradation 117
Rinderpest 176
Rumination 184
Rural butcher system 173
Rural development 3
rural farmers 14

S

Salt accumulation 61
Savannization 224
Seasonal pruning 85
Semi-intensive system 197
skill acquisition 43
slaughter 169, 171, 172, 177, 179, 180
smallholder farmers 19
social determinants 145
Soil compaction 98
soil conservation 87
soil erosion 99, 100
soil fauna 98
Soil moisture 59, 64, 67, 68, 69, 70, 71, 72, 73
Soil temperature 91, 92, 93, 94
Soil texture 57
Soil topography 55
Sound cropping pattern 40
Subsistence agriculture 86
supply chain 112
Surface tension 65
sustainability 2, 3, 4, 5, 7, 9, 10, 11, 12, 13, 14, 15, 18, 19
sustainable dairy farming 150, 159
Sustainable economic growth 32
Sustainable food production 9
Sustainable poultry 188
sustainable resource management 118, 134
Sustainable system 2, 3

T

tilling 116
Total digestible nutrients (TDN) 157
transpiration 96
trypanosome 177, 178
trypanosomiasis 176, 178
Trypanosomiasis 177

U

urban agriculture 108, 109, 111,

112, 114, 115, 116, 118, 120, 121, 122, 123, 124, 125, 126, 127, 128, 129, 130, 131, 133, 134, 135, 136, 137, 138, 139, 140
Urban agriculture 109, 112, 114, 117, 119, 123, 127, 134, 136
Urban Agriculture Working Group (UAWG) 120
urban communities 145
urban farming 108, 109, 111, 114, 115, 116, 120, 122, 125, 127, 133, 138
urbanization 109, 110
urban organic solid waste (UOSW) 139

V

Vapor diffusion 94
Vegetable mulches 94
vegetation 90
Vertical integration 236, 237
Veterinary inspection 172
vulnerability 131

W

wage employment 133
water management 136
Water pollution 119
weeding 87, 94, 99
weeds 96, 101